Prognostics and Health Management for Intelligent Electromechanical Systems

Hui Liu · Fang Cheng · Yanfei Li

Prognostics and Health Management for Intelligent Electromechanical Systems

Science Press
Beijing

Springer

Hui Liu
School of Traffic and Transportation Engineering
Central South University
Changsha, Hunan, China

Fang Cheng
School of Traffic and Transportation Engineering
Central South University
Changsha, Hunan, China

Yanfei Li
School of Mechanical and Electrical Engineering
Hunan Agricultural University
Changsha, Hunan, China

ISBN 978-981-96-7217-2 ISBN 978-981-96-7218-9 (eBook)
https://doi.org/10.1007/978-981-96-7218-9

Jointly published with Science Press
The print edition is not for sale in China mainland. Customers from China mainland please order the print book from: Science Press.

This Springer imprint is published by the registered company Springer Nature Singapore Pte Ltd.
The registered company address is: 152 Beach Road, #21-01/04 Gateway East, Singapore 189721, Singapore

Preface

Intelligent electromechanical systems play a crucial role in modern industry, particularly in the fields of automation and intelligent manufacturing. Against the backdrop of smart cities and Industry 4.0, these systems must achieve higher levels of service optimization and energy management to meet rapidly changing industrial demands and stringent environmental standards. These challenges require electromechanical systems to not only possess high degrees of automation and intelligence but also to have capabilities in predictive maintenance and fault prognostics.

The needs have spurred the development of predictive health management technologies. These technologies, by integrating advanced sensors, data analytics, and machine learning algorithms, enable real-time monitoring and analysis of the health status of intelligent electromechanical systems, predicting potential failures and performance degradation. This significantly enhances the reliability and safety of the equipment while extending its lifespan. Furthermore, it helps reduce maintenance costs and optimize resource allocation, playing a crucial role in driving the sustainable development and technological innovation of intelligent electromechanical systems.

As an interdisciplinary research field of data science and engineering, this book elaborates the research progress in predictive health management technologies for three advanced intelligent electromechanical systems: high-speed railway trains, robots, and new energy vehicles. It systematically describes the key technologies, including data acquisition, signal analysis, optimization models, ensemble learning, deep learning, and big data embedding. Specially, the analysis in the domain of high-speed trains focuses on systematic bearing fault diagnosis and remaining life prediction by analyzing bearing vibration signals in-depth, thus achieving precise monitoring and preventive maintenance of critical train components. The analysis in the domain of robots focuses on studying insulated gate bipolar transistors and metal-oxide-semiconductor field-effect transistors for remaining life prediction, constructing fault-tolerant control frameworks and dynamically adjusting operational strategies. The analysis in the domain of the new energy vehicles focuses on predicting the remaining life of capacitors and power supplies to optimize the operational efficiency and reliability of energy storage management systems, aiding in the sustainable development and intelligent upgrading of new energy vehicles.

This book is an essential reference across disciplines such as transportation engineering, mechanical engineering, automation, artificial intelligence, data science, and sustainable energy management. It offers practical insights and guidance for advancing predictive health management technologies in intelligent electromechanical systems, thereby enhancing industrial intelligence, boosting production efficiency, and ensuring comprehensive equipment safety.

The authors would like to express their sincere gratitude for the extensive experimental verification and other efforts undertaken by the team members.

Changsha, China
January 2025

Prof. Dr. -Ing. habil. Hui Liu

Contents

Nomenclature

AE	Auto-Encoder
ANN	Artificial Neural Network
ARIMA	Auto-Regressive Integrated Moving Average
BA	Bat Algorithm
CEEMD	Complete Ensemble Empirical Mode Decomposition
CF	Crest Factor
CFT	Continuous Fourier Transform
CLTC	China Light Vehicle Test Cycle
CNN	Convolutional Neural Network
CWT	Continuous Wavelet Transform
DBN	Deep Belief Networks
DCP	Distributed Computing Processing
DFT	Discrete Fourier Transform
DLM	Data Lifecycle Management
DT	Decision Tree Regression
DWT	Discrete Wavelet Transform
ECWT	Energy Concentration Wavelet Transform
EEMD	Ensemble Empirical Mode Decomposition
EEn	Energy Entropy
EMD	Empirical Mode Decomposition
ESS	Energy Storage Systems
EWT	Empirical Wavelet Transform
FFT	Fast Fourier Transform
FN	False Negative
FP	False Positive
GA	Genetic Algorithms
GAN	Generative Adversarial Network
GBT	Gradient-Boosted Trees
GPR	Gaussian Process Regression
GRU	Gated Recurrent Unit
GWO	Grey Wolf Optimizer

HDFS	Hadoop Distributed File System
HHT	Hilbert-Huang Transform
IES	Intelligent Electromechanical Systems
IGBT	Insulated Gate Bipolar Transistor
IMFs	Intrinsic Mode Functions
IoT	Internet of Things
KGE	Kling-Gupta Efficiency
KNN	K-Nearest Neighbors
Kur	Kurtosis
LR	Linear Regression
LSSVM	Least Squares Support Vector Machine
LSTM	Long Short-Term Memory
macro_F1	Macro F1 Score
macro_P	Macro-Average Precision
macro_R	Macro-Average Recall
MAE	Mean Absolute Error
MAPE	Mean Squared Error
MLP	Multilayer Perceptron
MOGWO	Multi-Objective Grey Wolf Optimizer
MSSC	Multi-Scale Segment Correlation
NEDC	New European Driving Cycle
NFTSM	Nonsingular Fast Terminal Sliding Mode
OCSSA	Osprey-Cauchy-Sparrow search algorithm
OCV	Open-Circuit Voltage
P	Pearson Correlation
PFP	Potential Failure Precursor
PHM	Prognostics and Health Management
PMSM	Permanent Magnet Synchronous Motor
PreMSSC	Predictive Multi-Scale Segment Correlation
RevIN	Reversible Instance Normalisation
RF	Random Forest
RMS	Rooted Mean Sqrt
RMSE	Root Mean Square Error
RNN	Recurrent Neural Network
RUL	Remaining Useful Life
RVM	Relevance Vector Machine
SampEn	Sample Entropy
SMC	Sliding Mode Control
SNR	Signal-to-Noise Ratio
SOC	State of Charge
SOH	State of Health
SpecEn	Spectral Entropy
STFT	Short Time Fourier Transformation
SVM	Support Vector Machine
TCN	Temporal Convolutional Network

TN	True Negative
TP	True Positive
VMD	Variational Mode Decomposition
WDT	Wavelet Decomposition Transform
WPD	Wavelet Packet Decomposition
WPT	Wavelet Packet Transform
WT	Wavelet Transform

List of Figures

List of Tables

Chapter 1
Introduction

Driven by the Industrial Revolution 4.0, intelligent electromechanical systems (IES) have become indispensable in modern industry and technology. The cross-domain coupling inherent in intelligent mechatronic systems makes it challenging to account for all interactions between components from different domains fully. Consequently, as system functions increase and structures become more complex, the potential failure modes also rise, placing higher demands on maintenance strategies.

The health status of IES is closely linked to their safe and reliable operation. Traditional preventive maintenance strategies have generally evolved through four stages: reactive maintenance, preventive maintenance, production-based maintenance, and condition-based maintenance. Traditional maintenance is primarily preventive, where maintenance is scheduled regardless of whether a fault has occurred, or reactive, where post-fault diagnostics and repairs are performed. Both preventive and reactive maintenance have unavoidable drawbacks. Recently, the introduction of Prognostics and Health Management (PHM) systems offers technical support to address these challenges, shifting from traditional fault diagnosis techniques to predictive technologies based on intelligent systems [20, 25]. PHM techniques start by gathering sensor data from machinery, then employ a range of intelligent reasoning models to evaluate the equipment's current health condition. These methods aim to anticipate potential failures and estimate the equipment's remaining useful life (RUL) by analyzing degradation patterns and additional data. This process supports informed decision-making and the planning of maintenance tasks. This book focuses on PHM for IES, with the comprehensive framework presented in Fig. 1.1.

H. Liu et al., *Prognostics and Health Management for Intelligent Electromechanical Systems*, https://doi.org/10.1007/978-981-96-7218-9_1

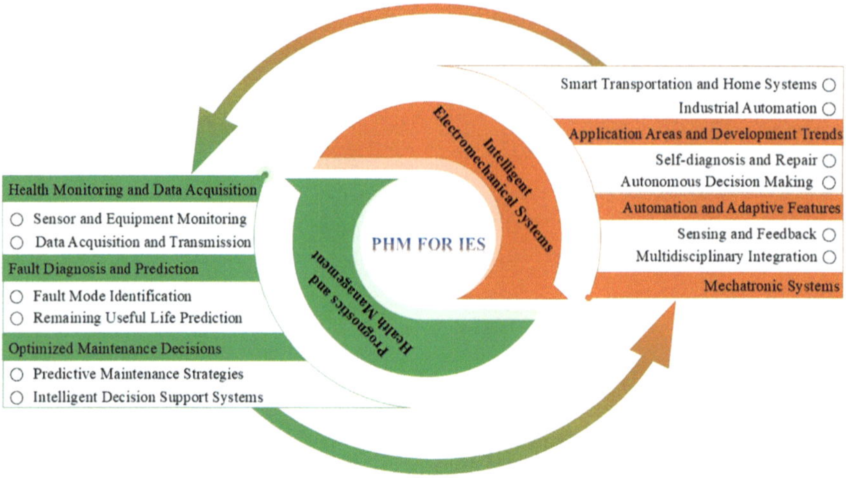

Fig. 1.1 Core of prognostics and health management for intelligent electromechanical systems in this book

1.1 Overview of Intelligent Electromechanical System

IES represent a sophisticated integration of mechanical components, electronics, and computational intelligence, designed to enhance the functionality and efficiency of diverse applications. These systems play a pivotal role in enabling PHM by providing critical real-time data and computational analysis [18]. The deployment of sensors and advanced diagnostics within IES facilitates the early detection of system anomalies and potential failures, which are essential for implementing effective PHM strategies [9]. Consequently, PHM leverages this data to enhance the reliability, maintenance efficiency, and safety of IES, ensuring timely interventions and reducing downtime through predictive analytics. IES are widely used across various fields, including transportation, robots, and energy, where their ability to intelligently respond and adapt to variables is crucial, as shown in Fig. 1.2. This book focuses on the application of IES in high-speed trains, robotics, and new energy vehicles, which serve as quintessential examples of IES.

1.1.1 High-Speed Trains

In the past decade, the complexity, integration, and intelligence of high-speed rail equipment have significantly increased. The traditional maintenance model, which is based on scheduled interventions, can no longer meet the evolving demands of maintenance and management [7]. Ongoing technological advancements have introduced a significant challenge: effectively evaluating the health status of rail transit

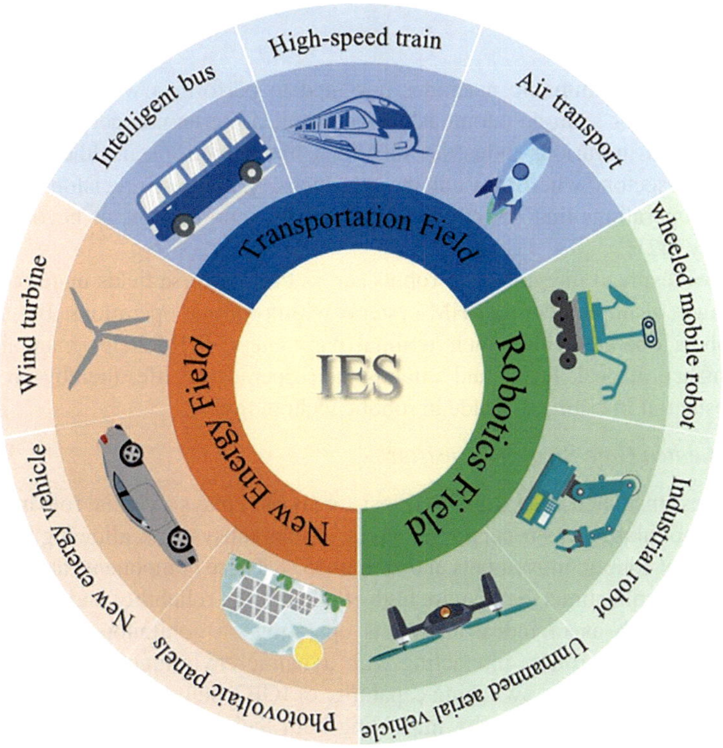

Fig. 1.2 Intelligent electromechanical systems

equipment by gathering, processing, and analyzing condition data, and making accurate, timely predictions about potential equipment failures. This issue has become a pressing problem that requires urgent attention.

As a critical component of the mechanical systems in train bogies, bearings play a vital role in supporting and transferring loads [29]. Bearings in train bogies must endure high rotational speeds and substantial loads while maintaining both precision and reliability. To meet these demands, advanced materials and innovative designs are employed to reduce friction and wear, thereby enhancing their durability and performance. These bearings are crucial for ensuring the smooth and safe operation of trains, as failure can lead to severe operational disruptions and safety hazards. Moreover, high-speed railway bearings are increasingly equipped with sensors for condition monitoring and predictive maintenance, integrating seamlessly into IES. This integration of technology facilitates the early identification of potential failures, enabling timely maintenance actions essential for the uninterrupted and efficient functioning of high-speed trains [95]. Considering the critical role bearings play in train safety and performance, it is vital to investigate dependable techniques for diagnosing bearing failures.

1.1.2 Robots

Robots are programmable machines designed to perform a wide range of tasks autonomously or semi-autonomously, often replicating human actions [32]. They are employed in various industries including manufacturing, healthcare, defense, and service sectors, where they enhance productivity, increase precision, and ensure safety in operations that are otherwise hazardous, monotonous, or beyond human capability [42].

The widespread integration of robots across these diverse fields underscores the importance of implementing PHM systems within robotic operations. By ensuring that robots operate within their optimal parameters, PHM helps maintain their performance at peak levels and extends their operational life, thereby protecting the substantial investments made in robotic technology.

(1) *Insulated Gate Bipolar Transistor*

In robotics, Insulated Gate Bipolar Transistor (IGBT) plays a crucial role in controlling and modulating electrical power across various subsystems, allowing for precise regulation of robotic movements and actions [57]. These components are especially valuable in applications that require high efficiency and reliability under demanding conditions, such as in industrial robots operating on continuous manufacturing lines or in service robots interacting with dynamic environmental factors. Moreover, the efficient power switching capability of IGBT helps reduce heat generation within robotic systems, thereby improving both their durability and performance [39]. Overall, employing IGBT in robots significantly enhances the development of advanced systems.

(2) *Metal–Oxide–Semiconductor Field-Effect Transistors*

Metal–Oxide–Semiconductor Field-Effect Transistors (MOSFET) is primarily used for power management and motor control in robotic applications. They enable precise control over the speed, direction, and torque of robotic motors, which is essential for tasks that require high precision, such as assembly work, surgical operations, or the operation of autonomous vehicles [10]. Their high switching speed and efficiency facilitate rapid and responsive control of actuators, which is crucial for dynamic and complex movements in robotics. Furthermore, MOSFET is known for its high efficiency and durability under high current loads, which helps minimize energy waste and reduces heat generation in robotic systems. This contributes to the longevity and reliability of robots, ensuring consistent performance even under demanding operational conditions. The robustness and adaptability of MOSFET make it an indispensable component in the design and functionality of modern robotic technologies.

1.1.3 New Energy Vehicles

With the increasing severity of global environmental pollution and the energy crisis, energy and environmental protection have become focal points of global attention [28]. Hence, new energy vehicles have emerged as a strategic response for countries to address climate change and reduce carbon emissions. New energy vehicles integrate advanced electric drive systems, energy management systems, and intelligent control technologies, achieving zero or low emissions through the efficient collaboration of core components such as batteries and electric motors.

However, the complexity and high demands on key components make PHM particularly important in this field [31]. PHM technology facilitates the real-time monitoring and predictive analysis of essential components, including batteries, motors, and power electronics, allowing for early detection of potential faults and optimization of maintenance schedules. This, in turn, enhances system reliability and safety. Therefore, the application of PHM technology in new energy vehicles is not only essential for improving vehicle operational efficiency but also a critical safeguard to ensure long-term, stable operation while maintaining high efficiency and environmental performance.

(1) *Capacitor*

The Capacitor is a crucial component responsible for efficiently storing and releasing energy to support the power system and energy recovery in new energy vehicles. They are capable of providing and absorbing large amounts of electrical energy in a short period, playing a key role in acceleration, climbing, and braking [73]. Capacitors are used in the regenerative braking system of new energy vehicles. When the vehicle decelerates or brakes, the motor operates as a generator, converting kinetic energy into electrical energy. The capacitors quickly store this energy and release it rapidly when acceleration is needed, enhancing energy efficiency and reducing dependence on the battery. Moreover, various electronic auxiliary systems in new energy vehicles, such as the electronic stability program and anti-lock braking system, also rely on capacitors to meet their short-term high-energy demands. For these high-performance capacitors, PHM technology is particularly important. By monitoring the capacitor's condition in real-time and analyzing performance data, PHM technology can predict the RUL of the capacitors, which not only prolongs the life of the capacitors but also aids in preserving the overall health and efficiency of the new energy vehicle system, preventing sudden failures and costly repairs, and ensuring long-term, efficient, and safe operation of the vehicle.

(2) *Power supply*

The Battery offers advantages such as high energy density, extended lifespan, and substantial operating voltage, making them the 'heart' of new energy electric vehicles and supplying power to these vehicles [96]. However, with prolonged use, battery aging failures can lead to serious battery incidents. For instance, battery failures can cause a decline in power system performance, even directly leading to mission failure and increased costs. Improper management of lithium-ion batteries in electric

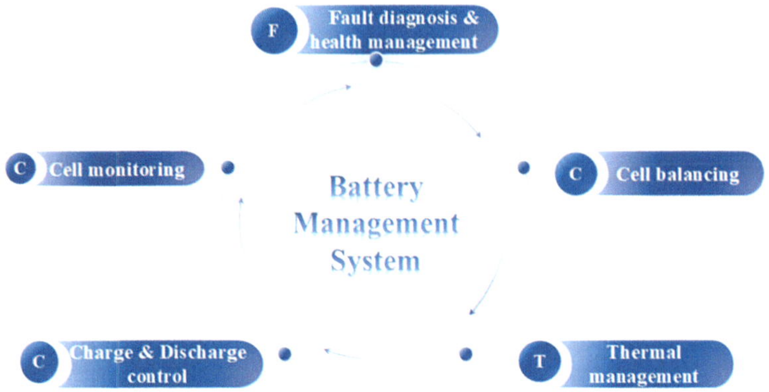

Fig. 1.3 Battery management system of new energy vehicles

vehicles can also result in fires or explosions. Overall, five main battery management systems are illustrated in Fig. 1.3. However, prematurely replacing the battery can lead to resource waste. Researching PHM methods for battery systems allows for more efficient utilization of these batteries, reducing unnecessary losses, and providing more stable and economical power support for new energy vehicles.

1.2 Research Status of Prognostics and Health Management in Intelligent Electromechanical System

The integration of PHM within IES has emerged as a critical research area, pivotal for enhancing the operational reliability and efficiency of contemporary technological applications. To comprehensively explore the development of IES, a literature review was conducted using the Web of Science database with 'PHM' as the search keyword. Based on the textual data downloaded from the Web of Science, it is revealed that the key topics in the PHM domain over the past five years predominantly include fault diagnosis and RUL.

Hence, this book predominantly focuses on these two critical aspects, which are indispensable for preventing unforeseen failures and optimizing maintenance schedules. Fault diagnosis processes within PHM frameworks are designed to detect and identify abnormalities in system operations at an early stage, effectively mitigating risks and minimizing downtime. Conversely, RUL prediction endeavors to accurately forecast the operational lifespan of system components, thereby facilitating preemptive maintenance strategies and ensuring sustainable system performance. The ongoing advancements in this domain not only extend the lifecycle of IES but also significantly boost their safety and reliability, which are vital in industries where precision and durability are of paramount importance.

1.2.1 Fault Diagnosis

Fault diagnosis is extensively applied to bearings in high-speed trains, with the predominant methods including acoustic fault diagnosis, axle temperature detection, and vibration fault diagnosis [88].

Acoustic fault diagnosis techniques primarily involve the installation of non-contact sensors on components like trackside locations, wheels, and axles. These sensors capture signals associated with the bearings as the train traverses the test zone, offering advantages such as non-invasive measurement, low cost, and strong early warning capabilities [26]. Specially, Li [36] Suggests a streamlined approach using a co-prime circular microphone array for diagnosing bearing faults and examines the acoustic traits of three fault signal types at varying rotation speeds. Wang [81] presents a new approach that utilizes the Grey Wolf Optimizer (GWO) and Support Vector Machine (SVM) to identify faults in wheelset bearings. The practical use of these methods is restricted because of the complex environment, significant interference from sound signals, and high distortion rates.

Temperature fault diagnosis involves assessing the state and operating conditions of bearings by measuring changes in their temperature during operation. When the temperature reaches a specified range, it can be used to infer bearing malfunctions [102]. Liu [49] developed a coupled thermo-mechanical model based on Newton's laws and the thermal grid method to study the variation of node and lubricating oil temperatures under different failure sizes for high-speed trains. Wang [79] created a model for the bearing rotor system to determine power loss and capture transient temperature profiles. However, during the initial stages of bearing faults such as wear, pitting, or cracking, there is no significant increase in heat generation. Only when the fault severity reaches a critical level does the bearing temperature exhibit substantial changes.

Vibration fault diagnosis is conducted using accelerometers to detect, and calculate the failure frequencies of components based on rotation speed and bearing dimensions, and compare these calculated frequencies with the frequency spectra of sampled signals for fault detection, localization, and severity assessment [71]. To overcome the limitations of insufficient diagnostic performance and low data quality commonly associated with Generative Adversarial Network (GAN)-based fault diagnosis approaches, Xu [86] introduced a dynamic model for bearings integrated with a multi-agent diverse GAN. Peng [61] developed a more robust one-dimensional convolutional neural network method based on one dimensional residual blocks, specifically designed to meet the fault detection requirements of high-speed train bearings in environments with high noise and variable load conditions. Currently, vibration analysis stands as the predominant technique for monitoring bearing conditions and diagnosing faults, providing benefits like non-invasiveness, affordability, and real-time capability.

1.2.2 Remaining Useful Life Prediction

RUL prediction is essential in the PHM of key components such as bearings, IGBT, MOSFET, capacitors, and batteries within intelligent electromechanical. The main approaches include physics-based models and data-driven models, as illustrated in Fig. 1.4.

(1) *High-speed trains*

The prediction of RUL for bearings in high-speed trains is mainly categorized into physical-based and data-driven approaches. Model-based approaches such as Hertz contact theory, L-P models, and finite element analysis methods [89] rely on expert prior knowledge. The complex and dynamic interplay of factors influencing bearing degradation makes pinpointing the precise degradation mechanism difficult, thereby constraining the predictive accuracy and utility of these models. On the other hand, data-driven approaches use vibration signal data from bearings collected by sensors to accurately extract degradation information using artificial intelligence models. These methods successfully correlate monitoring data with RUL values for detailed analysis.

(2) *Robots*

Significant research has been conducted on IGBT RUL prediction to identify degradation mechanisms and achieve accurate RUL estimates in the field of robots. The main degradation mechanism in IGBTs is adhesion degradation, alongside other processes like gate degradation. Direct methods, including X-ray and scanning acoustic microscopy analysis, can observe the level of degradation; however, these

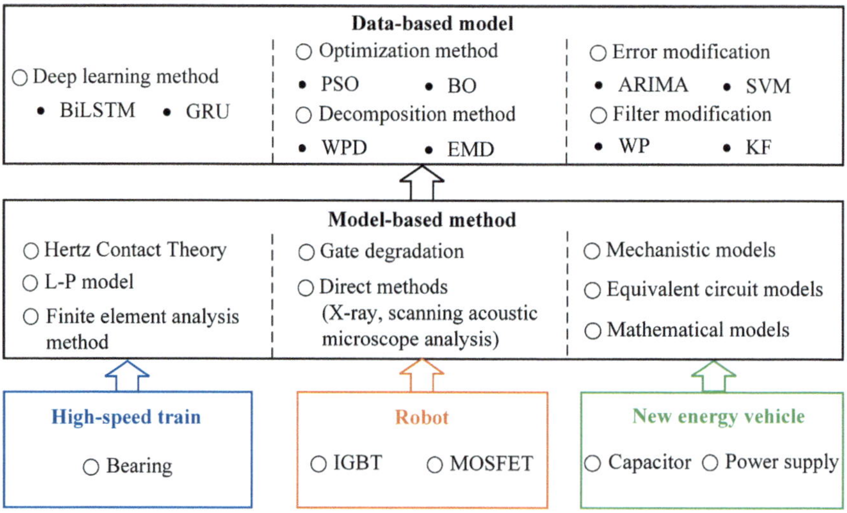

Fig. 1.4 RUL prediction for intelligent electromechanical systems

methods are offline, expensive, and destructive. Physical analysis methods based on failure mechanisms have been extensively developed, yet they struggle to account for dynamic condition changes and the presence of random noise in the system, resulting in significant errors in RUL estimation. Recently, data-driven approaches have evolved by tracking the progression of precursors to device degradation, utilizing parameters that define the aging of IGBTs to forecast [48, 82, 97].

MOSFET device characteristics are influenced by temperature, vibration, humidity, voltage, and other factors, and these characteristics change over time. The physics-based approach deeply explores the failure mechanisms of IGBTs, known for its clear physical importance and intuitive illustration. Yet, accurately applying Physics of Failure models [3] to real systems is complex, particularly when accounting for the interactions among various failure modes. Specifically, Hu et al. [24] created a 2D finite element model that assesses the stress performance and lifetime of the solder layer. Nevertheless, physics-based models are computationally demanding. Considerable research has been carried out on extracting data from sensor readings and employing it alongside data-driven methods to prevent and/or mitigate such failures.

(3) *New energy vehicles*

Capacitors are among the most critical components affecting the reliability of new energy vehicles. Predicting the RUL of capacitors is beneficial for assessing the health state in advance and devising reasonable maintenance strategies before system failures occur, thereby enhancing reliability and safety. Model-based strategies necessitate creating equivalent circuit model that mirrors the aging process, with life expectancy predicted through parameter identification. Recently, the rapid advancement of artificial intelligence has significantly benefited power electronics [74], with numerous machine learning-based methods being proposed for capacitor RUL prediction. These approaches do not rely on model assumptions and the data-driven models have proven more accurate than model-based techniques.

The typical power supply method for new energy vehicles, namely batteries, serves as the heart of these systems. RUL prediction typically falls into two categories: model-based and data-driven approaches. Physics-based models focus on the degradation processes and failure mechanisms, capturing the electrochemical reactions and physical changes that occur during charging and discharging cycles of lithium-ion batteries [15, 17, 21, 51, 77]. Despite these models offer high predictive accuracy and strong interpretability, their development is hindered by the complexities involved in modeling, time-consuming computations, challenges in data collection, and environmental fluctuations. On the other hand, data-driven methods, which derive RUL predictions for lithium-ion batteries by analyzing degradation data through various data analysis techniques, have gained significant traction [43].

1.3 Methodology of Prognostics and Health Management in Intelligent Electromechanical System

Considering that data-based models can be applied online, and are cost-effective and essentially non-destructive, these models have attracted widespread attention in PHM for IES. Currently, IES primarily incorporate three types of methods: (a) preprocessing—feature extraction model, (b) prediction model, and (c) postprocessing—error correction model, as illustrated in Fig. 1.5.

1.3.1 Feature Extraction Method

The feature extraction method effectively addresses the non-stationarity and nonlinearity of time series signals. Based on a comprehensive literature review, the mainstream feature extraction methods employed in PHM for IES can be further divided into three categories: (a) wavelet-based methods, (b) modal decomposition-based methods, and (c) deep learning-based methods.

(1) *High-speed trains*

To enhance the diagnostic accuracy of vibration signals from high-speed train bearings, data decomposition algorithms are commonly used. The feature vectors extracted are subsequently split into training and testing datasets, upon which classification algorithms are employed for identification. Common data decomposition methods include Wavelet Transform (WT) [92], Wavelet Packet Decomposition (WPD) [34], Empirical Wavelet Transform (EWT) [38], Ensemble Empirical Mode Decomposition (EEMD) [100], and Variational Mode Decomposition (VMD) [41]. Currently, popular deep learning models such as deep belief networks [37], stacked AutoEncoder (AE) [98], and Convolutional Neural Network (CNN) [61] are utilized to merge automatic feature extraction with type identification, fulfilling PHM tasks by harnessing deep hidden features embedded in high-dimensional data.

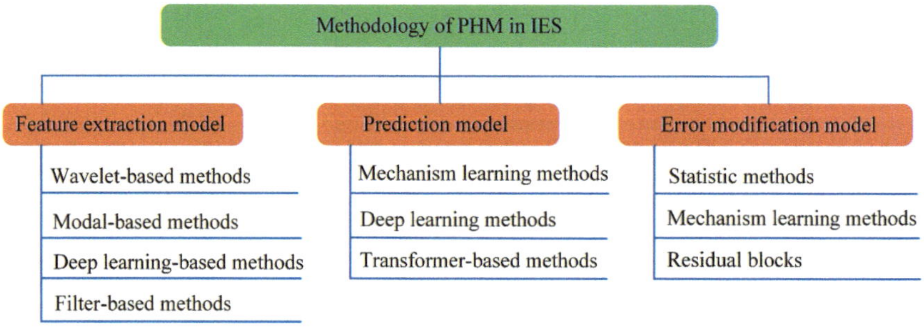

Fig. 1.5 Methodology of prognostics and health management in intelligent electromechanical system

(2) *Robots*

Power semiconductors (IGBTs and MOSFETs) are among the most critical components in modern power electronic circuits, especially in robotics research. In robotics, they operate under complex conditions and are prone to fatigue accumulation during repetitive operations, which can lead to malfunctions in the entire robotic system and cause significant economic losses. Given the non-linear, non-smooth, and complex time series characteristics of IGBT and MOSFET degradation, extensive research has employed EMD [53] and Complete Ensemble Empirical Mode Decomposition (CEEMD) [64] for noise reduction. Deng [11] and Ren [67] have effectively utilized Continuous Variational Mode Decomposition (SVMD) to process complex raw data during IGBT and MOSFET degradation, extracting multimodal features. Wang [82] employ six different methods for feature selection and fusion of IGBT characteristics.

(3) *New energy vehicles*

When batteries and capacitors coexist in energy storage systems (ESS) within new energy vehicles, energy management is crucial to ensure effective power distribution based on the advantages and disadvantages of each ESS, as well as safe operation. The presence of noise and capacity recovery phenomena in capacitor degradation curves makes it difficult to accurately trace their degradation trends through direct capacity prediction [4]. Qi [63] utilized VMD to process the aging data of capacitors, thereby removing noise interference and more effectively extracting capacity degradation features. Wang [80] employed a limited amount of training data to extract aging features using a CNN. Guo [21] processed the non-stationary signals of capacitor cycle life using Empirical Mode Decomposition (EMD).

Between discharge cycles, a battery's capacity may temporarily increase when not in use, a phenomenon referred to as capacity regeneration [46]. If not accounted for, this regeneration can significantly compromise the accuracy of RUL predictions when raw data is directly used [22]. Decomposition-based feature capture and data preprocessing methods have proven effective in addressing this challenge. Specifically, Ding [13] utilized the fast wavelet transform algorithm, combining Fixed Bandwidth Wavelet Decomposition (WDT) with a CNN to effectively capture signal characteristics across different frequency ranges and enhance computational efficiency. Building upon this segmentation strategy, Zhou [103], Li [40], and Chen [8] have successfully incorporated EMD with various predictive models to enhance prediction accuracy. Additionally, other adaptive decomposition methods widely used in lithium-ion batteries RUL prediction include EEMD [91], VMD [12], and CEEMDAN [52], among others.

1.3.2 Prediction Model

Prediction models are primarily classified into mechanism learning methods, deep learning methods, and transformer-based methods. Specifically for fault diagnosis

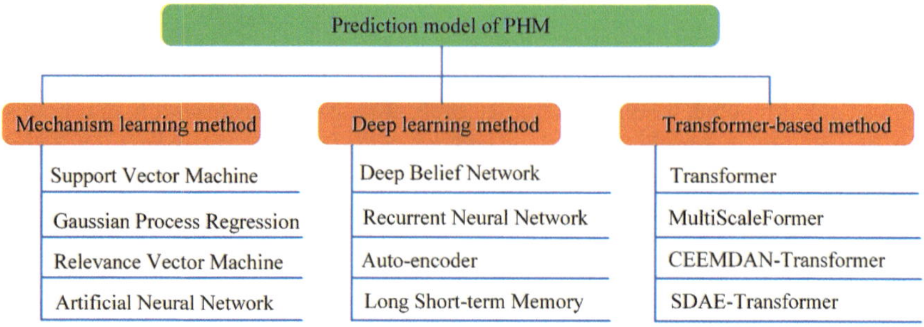

Fig. 1.6 Overview of prediction model in prognostics and health management

and RUL prediction of critical components in high-speed railways, robotics, and new energy vehicle domains, the summary of the book is depicted in Fig. 1.6.

(1) *High-speed trains*

Following the extraction and selection of features, high-speed trains axle box bearings' statistical parameters are processed through machine learning techniques including SVM [65], Gaussian Process Regression (GPR) [33], Artificial Neural Network (ANN) [90], and Relevance Vector Machine (RVM) [78], all of which have demonstrated efficacy in fault diagnosis and RUL prediction. With their superior capability for automatic feature recognition, deep learning methods are expected to revolutionize intelligent fault diagnosis and RUL prediction for bearings in high-speed railways. In recent developments, advanced neural networks such as Deep Belief Networks (DBN) [70], Long Short-Term Memory (LSTM) [45], and AE [54] have been successfully applied to RUL prediction, yielding competitive advantages. Embracing the attention mechanism, Tang [75] crafted a specialized Transformer model for identifying faults in bearings under fluctuating conditions, while Ding [14] has innovated a Transformer that integrates time–frequency analysis for fault detection in rotational machinery.

(2) *Robots*

For IGBT RUL prediction, techniques such as SVM [27] and Least Squares SVM (LSSVM) [98] are frequently utilized in machine learning algorithms for prediction purposes. Liu [48] presented a combined CNN-LSTM model to develop an IGBT lifetime prediction model using the collector-emitter voltage. Furthermore, He [23] proposed an improved Transformer variant, MultiScaleFormer, to constructs predictive model for the degradation of the IGBT gate oxide layer.

For MOSFET RUL prediction, various algorithms are employed to suppress noise, including the Kalman Filter, Particle Filter, RVM, Least Squares, and Extended Kalman Particle Filter algorithm [62]. Moreover, given that the entire lifecycle of a MOSFET can be segmented into different aging zones using the proposed aging feature, a segmented variable time-step metabolic GM(1,1) prediction method is introduced to further mitigate data series nonlinearity [93]. Kahraman [30] effectively captured the nonlinear degradation behavior of MOSFET by combining random forest and Bayesian algorithms, accurately predicting the RUL of MOSFET. Despite limited research on IGBT/MOSFET RUL prediction in robotics, existing studies establish a foundation for its application in this field.

(3) *New energy vehicles*

For capacitors RUL prediction, Zhou [104] demonstrated assessment of the RUL of supercapacitors used in vehicle electric motors through a hybrid method combining Genetic Algorithms (GA) and Recurrent Neural Network (RNN). Rigamonti [68] introduced a degradation indicator that is independent of measurement temperature and subsequently utilized a PF. Shahraki [69] proposed a framework that initially identifies the degradation point and then employs an LSTM model to predict the capacitors RUL. Luo [50] suggested a dual-stage online RUL prediction framework for capacitors, utilizing the BiLSTM network and H-infinity observer.

For battery RUL prediction, machine learning algorithms offer substantial advantages in terms of time efficiency, such as RVM [56], SVM [59], ELM [66], and so on. However, these algorithms have a limited capacity to learn nonlinear features. Based on deep learning models, LSTM [58], Gated Recurrent Unit (GRU) [22], AM with a Mixture of Experts [6] are used to predict lithium-ion batteries RUL under diverse charging strategies and conditions. In recent years, Transformer [5] and its variants, namely CEEMDAN-Transformer [83], and Stacked Denoising AE-Transformer [99], have made substantial progress due to their powerful representational capabilities.

1.3.3 Error Modification Model

Error modification is the most commonly used post-processing method in PHM for IES. It involves establishing a predictor to fit the forecast residual errors and applying the trained residual error prediction model to the test dataset. The error modification model leverages a subordinate predictor to learn the trends of errors and corrects the initial forecast results obtained by the primary predictor.

(1) *High-speed trains*

To improve the diagnostic precision of fault diagnosis tasks for high-speed train wheel bearings in noisy environments, error modification has been integrated into the PHM of IES, effectively preventing overfitting. Specifically, Peng [60] introduced a deeper one-dimensional CNN, which incorporates residual learning to counteract performance degradation in deep network models during training. Ni [55] proposed the Pulse-Coupled Integrated Residual Network, which tackles fault diagnosis across different operational conditions by deeply learning physical information. Additionally, Wang [78] utilized multi-layer improved residual shrinkage blocks to diminish the impact of noise, thus enhancing the precision of multi-class fault identification.

(2) *Robots*

To develop rapid and accurate technologies for practical application in IGBT fault diagnosis, Ahmed [1] proposed a multiclass SVM based on multinomial logistic regression, which has been proven highly effective in diagnosing the health conditions of modular multilevel converters. On the other hand, to address the limitations of traditional models, which lacks the implementation of compensation techniques for model correction and optimization. Ren [67] combines LSTM methods and error modification to refine and enhance the MOSFET RUL model. The above work provides insights into the development of the error modification model in IGBT/MOSFET within the field of robotics.

(3) *New energy vehicles*

To improve the precision in forecasting the RUL of capacitors for new energy vehicles, thereby ensuring the reliability of the entire capacitor bank, the error modification data post-processing method proves beneficial. Recently, Liu [44] incorporated a residual block temporal convolutional network, significantly enhancing the stability and accuracy of model predictions. The complex characteristics of lithium-ion batteries' capacity degradation sequences often challenge single data-driven models to fully capture their changes. For instance, Ding [13] developed an adaptive multiple error correction method that utilizes AutoRegressive Integrated Moving Average (ARIMA) to predict the residuals of the WDT-CNN model. This approach has demonstrated substantial enhancements in predicting the RUL of lithium-ion batteries.

1.4 The Necessity of Big Data Embedding in Prognostics and Health Management for Intelligent Electromechanical Systems

Progress in computer technology and the rise of big data have empowered deep learning to secure significant achievements in the PHM of IES. This technology can enhance data processing capabilities, improve fault prediction accuracy, achieve system optimization and adaptive adjustment, and facilitate cross-domain data integration and collaboration. For further details, see Fig. 1.7.

(1) *High-speed trains*

The immediate acquisition and examination of bearing data from high-speed trains allows for accurate equipment status monitoring. Yet, challenges in feature extraction arise due to the 'curse of dimensionality' associated with excessive monitoring data. Deep learning algorithms, such as CNN, have been widely applied to feature extraction in big data. Dong [16] employed an enhanced CNN to extract deep vibration features from rolling bearings and combined this with a geometric and statistical joint alignment method to obtain a projected shared feature space, facilitating the transfer of knowledge about bearing degradation processes and enabling degradation

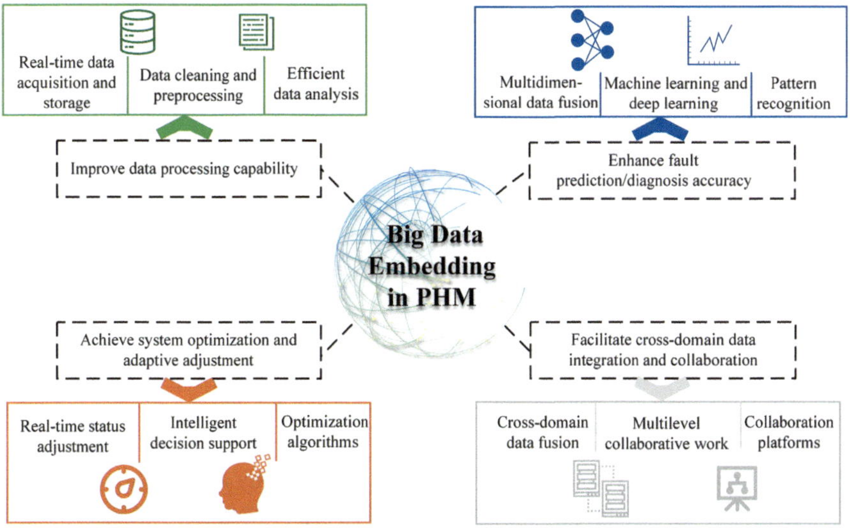

Fig. 1.7 Necessity of big data embedding in PHM

assessment under different operating conditions. Moreover, to overcome the loss of useful information that traditional feature extraction methods may cause, making it difficult to reflect the actual operational status of bearings in big data [105]. Shen [72] launched a deep learning strategy grounded in physics that begins by determining bearing health levels through a threshold model, then utilizes a CNN to autonomously derive high-level features essential for detecting faults. Furthermore, to address challenges such as the scarcity of failure samples and susceptibility to noise and monitoring errors in big data of rolling bearings for on-site monitoring, Lei [35] proposed a deep transfer learning method based on the joint generalized slice Wasserstein distance. Tang [76] introduced a cross-condition bearing diagnosis approach, leveraging the Kolmogorov-Arnold representation theorem.

(2) *Robots*

To cope with the constrained computational and storage capacities of robotic controllers, the computation of IGBT lifetime is performed on a big data platform. Liu [47] incorporated a dedicated big data server for the IGBT accelerated aging test, generating IGBT lifetime predictions based on junction temperature estimates derived from neural network and thermal network techniques.

The high-frequency sampling required for MOSFETs necessitates efficient storage, transformation, and analysis of massive amounts of experimental data within the prediction model. González [2] proposed a big data architecture in a distributed environment, see Fig. 1.8. This architecture uses Apache Flume and Apache Kafka to import data, employs Apache Spark for batch processing and real-time data handling, utilizes Apache Cassandra for batch calculations based on least squares, loading MOSFET RUL prediction results for further analysis and processing.

Fig. 1.8 Big data platform

(3) *New energy vehicles*

Big data-based energy management demonstrates significant potential for optimizing regional new energy vehicle networks and power systems. Implementing this approach enables more effective coordination of vehicle operations, optimized power distribution, and reduction of energy wastage from diesel, battery, and supercapacitor sources [85]. Recently, Zhao [101] introduced a big data statistical analysis for fault diagnosis, utilizing local outlier factors and clustering outlier diagnosis algorithms to validate detection results. Similarly, Dubarry [19] used big data to analyze the aging mechanisms of various battery chemistries, employing Independent Component Analysis to delineate the laws governing battery aging and explicitly uncover the contributions of Lithium Loss Inventory and Lithium Active Mass to overall capacity decline. In a practical application, Wang [84] developed a data-driven methodology based on extensive real-world data from new energy vehicles to diagnose battery charging capacity issues. This approach employs statistical analysis of large datasets to identify anomalies in battery charging capacity. Additionally, Yu [94] explored the operational mechanisms of a Big-Data-Based power battery recovery platform, demonstrating how integrated big-data attributes can enhance system functionalities. Furthermore, Xue [87] investigated a new method for diagnosing and detecting anomalies in electric scooter battery packs stored on cloud platforms. This method calculates a diagnostic coefficient based on a Gaussian distribution, enhancing the ability to promptly detect and pinpoint faulty batteries, thus ensuring timely intervention and maintenance.

1.5 Scope of the Book

This book explores the critical role of PHM technologies, focusing on their application in three key domains of IES: high-speed railways, robotics, and new energy vehicles. By advancing data processing algorithms, fault diagnosis, and predictive maintenance strategies, the book illustrates how to implement PHM in order to

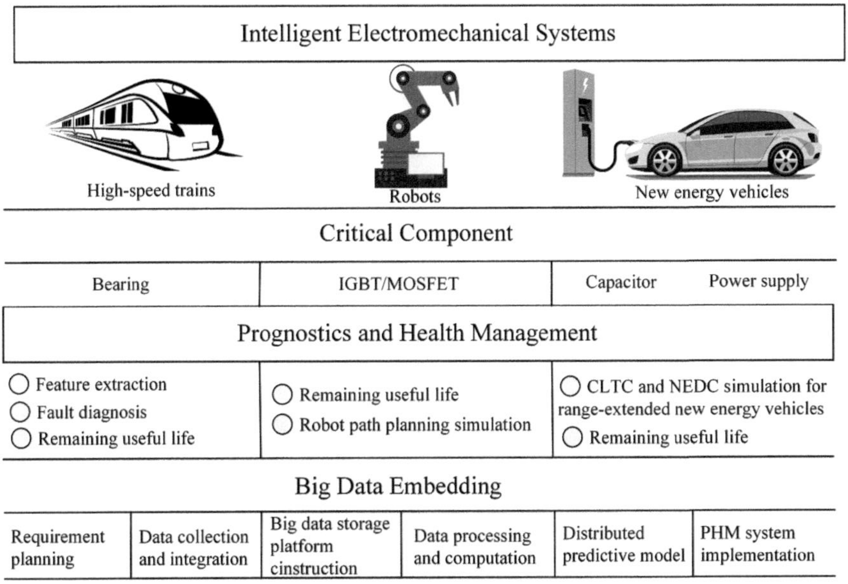

Fig. 1.9 The framework of the book

enhance system reliability and efficiency across these four sectors. The structure of the book is illustrated in Fig. 1.9.

- The first section is dedicated to high-speed trains, providing a detailed discussion of several key technological advancements. It begins by exploring ensemble intelligent diagnosis for bearing faults, a method that integrates multiple diagnostic models. Subsequently, the section delves into feature extraction, outlining techniques used to analyze and distill meaningful data from complex vibration patterns. Additionally, the section concludes with an examination of bearing RUL prediction based on deep learning methods.
- The second section addresses the prediction of RUL for IGBT and MOSFET components, which are crucial for the power electronics used in robotics. It first introduces an optimization-based RUL prediction for IGBT, which utilizes optimization algorithms to enhance the accuracy. Following this, this section examines a decomposition-based prediction for MOSFET RUL, where decomposition techniques are employed to analyze and predict the life cycle of MOSFETs. Subsequently, these works are applied to the fault-tolerant control of wheeled mobile robots.
- The third part of the document delves into new energy vehicles, focusing on two specific applications concerning capacitors and power supplies. Initially, it discusses predictions for the remaining life of capacitors using an error-modification-based approach. This method adjusts predictive models to enhance accuracy in estimating how long capacitors will continue to function effectively. Subsequently, the section employs joint simulation tools from Cruise and

MATLAB to conduct simulations under CLTC and NEDC conditions for range-extended new energy vehicles. These simulations provide detailed insights into battery capacity degradation. Finally, it explores filter-modification-based predictions for the RUL of power supplies in new energy vehicles, offering strategies to extend the longevity and reliability of these essential components.

- The fourth part focuses on PHM for IES. It discusses the construction of a big-data storage platform designed to handle extensive data generated by these systems. Additionally, this section introduces a distributed predictive model tailored for electromechanical systems, which leverages the scalability and flexibility of distributed computing to enhance predictive accuracy and system responsiveness. These methods exemplify the cutting-edge predictive health management techniques, vital for optimizing the reliability and efficiency of IES.

This book not only offers engineers and technicians guidance on both theory and practical operations but also provides researchers with a platform to explore the latest advancements in PHM technologies.

References

1. Ahmed HOA, Yu YX, Wang QH, Darwish M, Nandi AK (2022) Intelligent fault diagnosis framework for modular multilevel converters in HVDC transmission. Sensors 22(1):362. https://doi.org/10.3390/s22010362
2. Alonso-González CJ, Pulido B, Cartón M, Bregon A (2019) A big data architecture for fault prognostics of electronic devices: application to power MOSFETs. IEEE Access 7:102160–102173. https://doi.org/10.1109/access.2019.2929111
3. Ceccarelli L, Kotecha RM, Bahman AS, Iannuzzo F, Mantooth HA (2019) Mission-profile-based lifetime prediction for a SiC MOSFET power module using a multi-step condition-mapping simulation strategy. IEEE Trans Power Electron 34(10):9698–9708
4. Chaari R, Briat O, Vinassa JM (2014) Capacitance recovery analysis and modelling of supercapacitors during cycling ageing tests. Energy Convers Manage 82:37–45. https://doi.org/10.1016/j.enconman.2014.02.051
5. Chen DQ, Hong WC, Zhou XZ (2022) Transformer network for remaining useful life prediction of lithium-ion batteries. IEEE Access 10:19621–19628. https://doi.org/10.1109/access.2022.3151975
6. Chen DQ, Zhou XZ (2024) AttMoE: attention with mixture of experts for remaining useful life prediction of lithium-ion batteries. J Energy Storage 84:110780. https://doi.org/10.1016/j.est.2024.110780
7. Chen HT, Jiang B (2020) A review of fault detection and diagnosis for the traction system in high-speed trains. IEEE Trans Intell Transp Syst 21(2):450–465. https://doi.org/10.1109/tits.2019.2897583
8. Chen JC, Chen TL, Liu WJ, Cheng CC, Li MG (2021) Combining empirical mode decomposition and deep recurrent neural networks for predictive maintenance of lithium-ion battery. Adv Eng Inf 50:101405. https://doi.org/10.1016/j.aei.2021.101405
9. Dai N, Lei IM, Li ZY, Li Y, Fang P, Zhong JW (2023) Recent advances in wearable electromechanical sensors—moving towards machine learning-assisted wearable sensing systems. Nano Energy 105:108041. https://doi.org/10.1016/j.nanoen.2022.108041
10. Das A, Rewari S, Kanaujia BK, Gupta RS (2022) Recent technological advancement in surrounding gate MOSFET for biosensing applications—a synoptic study. SILICON 14(10):5133–5143. https://doi.org/10.1007/s12633-021-01288-w

11. Deng KT, Xu XL, Yuan F, Zhang TY, Xu YL, Xie TZ, Song YQ, Zhao RQ (2024) An analytical approach for IGBT life prediction using successive variational mode decomposition and bidirectional long short-term memory networks. Electronics 13(20):4002. https://doi.org/10.3390/electronics13204002
12. Ding GR, Wang WB, Zhu T (2022) Remaining useful life prediction for lithium-ion batteries based on CS-VMD and GRU. IEEE Access 10:89402–89413. https://doi.org/10.1109/access.2022.3167759
13. Ding P, Liu XJ, Li H, Huang Z, Zhang K, Shao L, Abedinia O (2021) Useful life prediction based on wavelet packet decomposition and two-dimensional convolutional neural network for lithium-ion batteries. Renew Sustain Energy Rev 148:111287. https://doi.org/10.1016/j.rser.2021.111287
14. Ding YF, Jia MP, Miao QH, Cao YD (2022) A novel time-frequency transformer based on self-attention mechanism and its application in fault diagnosis of rolling bearings. Mech Syst Sig Process 168:108616. https://doi.org/10.1016/j.ymssp.2021.108616
15. Dong GZ, Xu Y, Wei ZB (2022) A hierarchical approach for finite-time H-∞ state-of-charge observer and probabilistic lifetime prediction of lithium-ion batteries. IEEE Trans Energy Convers 37(1):718–728. https://doi.org/10.1109/tec.2021.3109896
16. Dong SZ, Wen GR, Lei ZH, Zhang ZF (2021) Transfer learning for bearing performance degradation assessment based on deep hierarchical features. ISA Trans 108:343–355. https://doi.org/10.1016/j.isatra.2020.09.004
17. Downey A, Lui YH, Hu C, Laflamme S, Hu S (2019) Physics-based prognostics of lithium-ion battery using non-linear least squares with dynamic bounds. Reliab Eng Syst Saf 182:1–12. https://doi.org/10.1016/j.ress.2018.09.018
18. Du RH, Xiao J, Chang SD, Zhao LC, Wei KX, Zhang WM, Zou HX (2023) Mechanical energy harvesting in traffic environment and its application in smart transportation. J Phys D-Appl Phys 56(37):373002. https://doi.org/10.1088/1361-6463/acdadb
19. Dubarry M, Beck D (2021) Analysis of synthetic voltage vs. capacity datasets for big data Li-ion diagnosis and prognosis. Energies 14(9):2371. https://doi.org/10.3390/en14092371
20. Gharib H, Kovács G (2023) A review of prognostic and health management (PHM) methods and limitations for marine diesel engines: new research directions. Machines 11(7):695. https://doi.org/10.3390/machines11070695
21. Guo F, Lv HT, Wu XW, Yuan XH, Liu LL, Ye JL, Wang T, Fu LJ, Wu YP (2023) A machine learning method for prediction of remaining useful life of supercapacitors with multi-stage modification. J Energy Storage 73:109160. https://doi.org/10.1016/j.est.2023.109160
22. Guo F, Wu XW, Liu LL, Ye JL, Wang T, Fu LJ, Wu YP (2023) Prediction of remaining useful life and state of health of lithium batteries based on time series feature and Savitzky-Golay filter combined with gated recurrent unit neural network. Energy 270:126880. https://doi.org/10.1016/j.energy.2023.126880
23. He SL, Yu ML, Chen YQ, Wei ZZ, Yu LB, Zhang C, Ni YH (2024) Prediction of IGBT gate oxide layer's performance degradation based on MultiScaleFormer network. Micromachines 15(8):985. https://doi.org/10.3390/mi15080985
24. Hu BR, Gonzalez JO, Ran L, Ren H, Zeng Z, Lai W, Gao B, Alatise O, Lu H, Bailey C, Mawby P (2017) Failure and reliability analysis of a SiC power module based on stress comparison to a Si device. IEEE Trans Device Mater Reliab 17(4):727–737. https://doi.org/10.1109/tdmr.2017.2766692
25. Huang C, Bu SQ, Lee HH, Chan KW, Yung WKC (2024) Prognostics and health management for induction machines: a comprehensive review. J Intell Manuf 35(3):937–962. https://doi.org/10.1007/s10845-023-02103-6
26. Huang HD, Liu F, Geng L, Liu YB, Ren ZH, Zhao YK, Lei XJ, Lu XY (2019) Fault diagnosis accuracy improvement using wayside rectangular microphone array for health monitoring of railway-vehicle wheel bearing. IEEE Access 7:87410–87424. https://doi.org/10.1109/access.2019.2924832
27. Huang HZ, Wang HK, Li YF, Zhang LL, Liu ZL (2015) Support vector machine based estimation of remaining useful life: current research status and future trends. J Mech Sci Technol 29:151–163. https://doi.org/10.1007/s12206-014-1222-z

28. Jiang F, Yuan XH, Hu LL, Xie GM, Zhang ZQ, Li XP, Hu J, Wang C, Wang HC (2024) A comprehensive review of energy storage technology development and application for pure electric vehicles. J Energy Storage 86:111159. https://doi.org/10.1016/j.est.2024.111159
29. Jiang WL, Wang CY, Zou JY, Zhang SQ (2021) Application of deep learning in fault diagnosis of rotating machinery. Processes 9(6):919. https://doi.org/10.3390/pr9060919
30. Kahraman CL, Roman D, Kirschbaum L, Flynn D, Swingler J (2024) Machine learning pipeline for power electronics state of health assessment and remaining useful life prediction. IEEE Access 12:136727–136749. https://doi.org/10.1109/ACCESS.2024.3460177
31. Kandidayeni M, Trovao JP, Soleymani M, Boulon L (2022) Towards health-aware energy management strategies in fuel cell hybrid electric vehicles: a review. Int J Hydrogen Energy 47(17):10021–10043. https://doi.org/10.1016/j.ijhydene.2022.01.064
32. Kumar P, Khalid S, Kim HS (2023) Prognostics and health management of rotating machinery of industrial robot with deep learning applications—a review. Mathematics 11(13):3008. https://doi.org/10.3390/math11133008
33. Kumar PS, Kumaraswamidhas LA, Laha SK (2021) Bearing degradation assessment and remaining useful life estimation based on Kullback-Leibler divergence and Gaussian processes regression. Measurement 174:108948. https://doi.org/10.1016/j.measurement.2020.108948
34. Laala W, Guedidi A, Guettaf A (2023) Bearing faults classification based on wavelet transform and artificial neural network. Int J Syst Assur Eng Manag 14(1):37–44. https://doi.org/10.1007/s13198-020-01039-x
35. Lei N, Cui JP, Han JC, Chen X, Tang YF (2024) Rolling bearing fault diagnosis using deep transfer learning based on joint generalized sliced Wasserstein distance. IEEE Access 12:41452–41463. https://doi.org/10.1109/access.2024.3375400
36. Li C, Chen CZ, Gu XJ (2023) Acoustic-based rolling bearing fault diagnosis using a co-prime circular microphone array. Sensors 23(6):3050. https://doi.org/10.3390/s23063050
37. Li HH, Wang HK, Xie ZW, He MQ (2022) Fault diagnosis of railway freight car wheelset based on deep belief network and cuckoo search algorithm. Proc Inst Mech Eng Part F-J Rail Rapid Transit 236(5):501–510. https://doi.org/10.1177/09544097211029155
38. Li JM, Wang H, Wang XD, Zhang YG (2020) Rolling bearing fault diagnosis based on improved adaptive parameterless empirical wavelet transform and sparse denoising. Measurement 152:107392. https://doi.org/10.1016/j.measurement.2019.107392
39. Li Q, Li YB, Fu HD, Tu CM, Xiao B, Xiao F, Zhai DY, Lu JW (2023) Review of the failure mechanism and methodologies of IGBT bonding wire. IEEE Trans Comp Packag Manuf Technol 13(7):1045–1057. https://doi.org/10.1109/tcpmt.2023.3297224
40. Li XY, Zhang L, Wang ZP, Dong P (2019) Remaining useful life prediction for lithium-ion batteries based on a hybrid model combining the long short-term memory and Elman neural networks. J Energy Storage 21:510–518. https://doi.org/10.1016/j.est.2018.12.011
41. Li ZP, Chen JL, Zi YY, Pan J (2017) Independence-oriented VMD to identify fault feature for wheel set bearing fault diagnosis of high speed locomotive. Mech Syst Sig Process 85:512–529. https://doi.org/10.1016/j.ymssp.2016.08.042
42. Lin ZH, Huang ZW, Yang SJ, Wu CH, Fang S, Liu ZC, Yang M, Xie ZH, Li DG, Zou YT (2024) Survey on task-centric robot battery management: a neural network framework. J Power Sources 610:234674. https://doi.org/10.1016/j.jpowsour.2024.234674
43. Lipu MSH, Abd Rahman MS, Mansor M, Rahman T, Ansari S, Fuad A, Hannan MA (2024) Data driven health and life prognosis management of supercapacitor and lithium-ion battery storage systems: developments, implementation aspects, limitations, and future directions. J Energy Storage 98:113172. https://doi.org/10.1016/j.est.2024.113172
44. Liu CL, Li DZ, Wang LC, Li LW, Wang K (2022) Strong robustness and high accuracy in predicting remaining useful life of supercapacitors. APL Mater 10(6):061106. https://doi.org/10.1063/5.0092074
45. Liu JQ, Pan CL, Lei FB, Hu DB, Zuo HF (2021) Fault prediction of bearings based on LSTM and statistical process analysis. Reliab Eng Syst Saf 214:107646. https://doi.org/10.1016/j.ress.2021.107646

46. Liu JZ, Liu XT (2023) An improved method of state of health prediction for lithium batteries considering different temperature. J Energy Storage 63:107028. https://doi.org/10.1016/j.est.2023.107028
47. Liu L, Tang L, Jiang HP, Wei FY, Li ZH, Du CH, Peng QL, Lu GC (2024) Reliability evaluation of IGBT power module on electric vehicle using big data. J Semicond 45(5):052301. https://doi.org/10.1088/1674-4926/45/5/052301
48. Liu QL, Tong QB, Wang L, An Guoping IGBT (2021) Life prediction based on CNN1D-LSTM hybrid model. International conference on electrical and information technologies for rail transportation. Springer, pp 660–668
49. Liu YQ, Wang BS, Yang SP, Liao YY, Guo T (2022) Characteristic analysis of mechanical thermal coupling model for bearing rotor system of high-speed train. Appl Math Mech-Engl Ed 43(9):1381–1398. https://doi.org/10.1007/s10483-022-2893-5
50. Lou GM, Lin WW, Huang GX, Xiang W (2023) A two-stage online remaining useful life prediction framework for supercapacitors based on the fusion of deep learning network and state estimation algorithm. Eng Appl Artif Intell 123:106399. https://doi.org/10.1016/j.engappai.2023.106399
51. Luo K, Zheng HR, Shi ZC (2023) A simple feature extraction method for estimating the whole life cycle state of health of lithium-ion batteries using transformer-based neural network. J Power Sources 576:233139. https://doi.org/10.1016/j.jpowsour.2023.233139
52. Luo TY, Liu M, Shi P, Duan GR, Cao XB (2023) A hybrid data preprocessing-based hierarchical attention BiLSTM network for remaining useful life prediction of spacecraft lithium-ion batteries. IEEE Trans Neural Networks Learn Syst 35(12):18076–18089. https://doi.org/10.1109/tnnls.2023.3311443
53. Manikandan R, Singh RR (2024) Fault diagnosis of wind turbine power converter using intrinsic mode functions with relative energy entropy. Circuit World 50(1):90–107. https://doi.org/10.1108/cw-09-2022-0241
54. Mao WT, Feng WS, Liu YM, Zhang D, Liang XH (2021) A new deep auto-encoder method with fusing discriminant information for bearing fault diagnosis. Mech Syst Sig Process 150:107233. https://doi.org/10.1016/j.ymssp.2020.107233
55. Ni Q, Ji JC, Halkon B, Feng K, Nandi AK (2023) Physics-Informed Residual Network (PIResNet) for rolling element bearing fault diagnostics. Mech Syst Sig Process 200:110544. https://doi.org/10.1016/j.ymssp.2023.110544
56. Nuhic A, Terzimehic T, Soczka-Guth T, Buchholz M, Dietmayer K (2013) Health diagnosis and remaining useful life prognostics of lithium-ion batteries using data-driven methods. J Power Sources 239:680–688. https://doi.org/10.1016/j.jpowsour.2012.11.146
57. Oh H, Han B, McCluskey P, Han C, Youn BD (2015) Physics-of-failure, condition monitoring, and prognostics of insulated gate bipolar transistor modules: a review. IEEE Trans Power Electron 30(5):2413–2426. https://doi.org/10.1109/tpel.2014.2346485
58. Park K, Choi Y, Choi WJ, Ryu HY, Kim H (2020) LSTM-based battery remaining useful life prediction with multi-channel charging profiles. IEEE Access 8:20786–20798. https://doi.org/10.1109/access.2020.2968939
59. Patil MA, Tagade P, Hariharan KS, Kolake SM, Song T, Yeo TJ, Doo S (2015) A novel multistage Support Vector Machine based approach for Li ion battery remaining useful life estimation. Appl Energy 159:285–297. https://doi.org/10.1016/j.apenergy.2015.08.119
60. Peng DD, Liu ZL, Wang H, Qin Y, Jia LM (2019) A novel deeper one-dimensional CNN with residual learning for fault diagnosis of wheelset bearings in high-speed trains. IEEE Access 7:10278–10293. https://doi.org/10.1109/access.2018.2888842
61. Peng DD, Wang H, Liu ZL, Zhang W, Zuo MJ, Chen J (2020) Multibranch and multiscale CNN for fault diagnosis of wheelset bearings under strong noise and variable load condition. IEEE Trans Ind Inf 16(7):4949–4960. https://doi.org/10.1109/tii.2020.2967557
62. Pugalenthi K, Park H, Raghavan N (2019) Prognosis of power MOSFET resistance degradation trend using artificial neural network approach. Microelectron Reliab 100:113467. https://doi.org/10.1016/j.microrel.2019.113467

63. Qi GH, Ma N, Wang K (2024) Predicting the remaining useful life of supercapacitors under different operating conditions. Energies 17(11):2585. https://doi.org/10.3390/en17112585
64. Qin N, Wang TW, Huang DQ, Zhang YM, Ma L (2023) An open-switch fault diagnosis method for single-phase PWM rectifier based on CEEMD-DNN. Int J Control Autom Syst 21(10):3430–3442. https://doi.org/10.1007/s12555-021-0905-3
65. Rai A, Upadhyay SH (2018) Intelligent bearing performance degradation assessment and remaining useful life prediction based on self-organising map and support vector regression. Proc Inst Mech Eng Part C-J Mech Eng Sci 232(6):1118–1132. https://doi.org/10.1177/0954406217700180
66. Razavi-Far R, Chakrabarti S, Saif M, Zio E (2019) An integrated imputation-prediction scheme for prognostics of battery data with missing observations. Expert Syst Appl 115:709–723. https://doi.org/10.1016/j.eswa.2018.08.033
67. Ren HY, Yu YY, Liu JL, Zhou JJ, Du X (2023) Lifetime prediction of power MOSFET based on LSTM with successive variational mode decomposition and error compensation. IEICE Electron Express 20(16):1–6. https://doi.org/10.1587/elex.20.20230277
68. Rigamonti M, Baraldi P, Zio E, Astigarraga D, Galarza A (2016) Particle filter-based prognostics for an electrolytic capacitor working in variable operating conditions. IEEE Trans Power Electron 31(2):1567–1575. https://doi.org/10.1109/tpel.2015.2418198
69. Shahraki AF, Al-Dahidi S, Taleqani AR, Yadav OP (2023) Using LSTM neural network to predict remaining useful life of electrolytic capacitors in dynamic operating conditions. Proc Inst Mech Eng Part O-J Risk Reliab 237(1):16–28. https://doi.org/10.1177/1748006x221087503
70. Shao HD, Jiang HK, Zhang HZ, Liang TC (2018) Electric locomotive bearing fault diagnosis using a novel convolutional deep belief network. IEEE Trans Ind Electron 65(3):2727–2736. https://doi.org/10.1109/tie.2017.2745473
71. Shao HD, Jiang HK, Zhao K, Wei DD, Li XQ (2018) A novel tracking deep wavelet auto-encoder method for intelligent fault diagnosis of electric locomotive bearings. Mech Syst Sig Process 110:193–209. https://doi.org/10.1016/j.ymssp.2018.03.011
72. Shen S, Lu H, Sadoughi M, Hu C, Nemani V, Thelen A, Webster K, Darr M, Sidon J, Kenny S (2021) A physics-informed deep learning approach for bearing fault detection. Eng Appl Artif Intell 103:104295. https://doi.org/10.1016/j.engappai.2021.104295
73. Sun JL, Luo BC, Li HX (2022) A review on the conventional capacitors, supercapacitors, and emerging hybrid ion capacitors: past, present, and future. Adv Energy Sustain Res 3(6):2100191. https://doi.org/10.1002/aesr.202100191
74. Sun Q, Yang LC, Li HS, Sun GD (2023) RUL prediction for AECs of power electronic systems based on machine learning and error compensation. J Intell Fuzzy Syst 44(5):7407–7417. https://doi.org/10.3233/jifs-220866
75. Tang J, Zheng GH, Wei C, Huang WB, Ding XX (2022) Signal-transformer: a robust and interpretable method for rotating machinery intelligent fault diagnosis under variable operating conditions. IEEE Trans Instrum Meas 71:1–11. https://doi.org/10.1109/TIM.2022.3169528
76. Tang ZY, Hou XH, Wang X, Zou JF (2024) A cross-working condition-bearing diagnosis method based on image fusion and a residual network incorporating the Kolmogorov-Arnold representation theorem. Appl Sci-Basel 14(16):7254. https://doi.org/10.3390/app14167254
77. Tian JQ, Xu RL, Wang YJ, Chen ZH (2021) Capacity attenuation mechanism modeling and health assessment of lithium-ion batteries. Energy 221:119682. https://doi.org/10.1016/j.energy.2020.119682
78. Wang B, Lei YG, Li NP, Li NB (2020) A hybrid prognostics approach for estimating remaining useful life of rolling element bearings. IEEE Trans Reliab 69(1):401–412. https://doi.org/10.1109/tr.2018.2882682
79. Wang BS, Liu YQ, Liao YY, Wang YX (2024) Analysis of temperature characteristics of high-speed train wheelset system under complex operating conditions. Int J Mech Syst Dynam 4(3):331–345. https://doi.org/10.1002/msd2.12122

80. Wang CX, Xiong R, Tian JP, Lu JH, Zhang CM (2022) Rapid ultracapacitor life prediction with a convolutional neural network. Appl Energy 305:117819. https://doi.org/10.1016/j.apenergy.2021.117819
81. Wang TH, Meng HY, Zhang F, Qin R (2024) Fault detection of wheelset bearings through vibration-sound fusion data based on grey wolf optimizer and support vector machine. Technologies 12(9):144. https://doi.org/10.3390/technologies12090144
82. Wang X, Zhou ZW, He SL, Liu JB, Cui W (2023) Performance degradation modeling and its prediction algorithm of an IGBT gate oxide layer based on a CNN-LSTM network. Micromachines 14(5):959. https://doi.org/10.3390/mi14050959
83. Wang ZL, Liu YL, Wang F, Wang H, Su M (2023) Capacity and remaining useful life prediction for lithium-ion batteries based on sequence decomposition and a deep-learning network. J Energy Storage 72:108085. https://doi.org/10.1016/j.est.2023.108085
84. Wang ZP, Song CB, Zhang L, Zhao Y, Liu P, Dorrell DG (2022) A data-driven method for battery charging capacity abnormality diagnosis in electric vehicle applications. IEEE Trans Transp Electrif 8(1):990–999. https://doi.org/10.1109/tte.2021.3117841
85. Wei ZB, Liu KL, Liu XH, Li Y, Du L, Gao F (2023) Multilevel data-driven battery management: from internal sensing to big data utilization. IEEE Trans Transp Electrif 9(4):4805–4823. https://doi.org/10.1109/tte.2023.3301990
86. Xu K, Kong XG, Wang QB, Han B, Sun LQ (2023) Intelligent fault diagnosis of bearings under small samples: a mechanism-data fusion approach. Eng Appl Artif Intell 126:107063. https://doi.org/10.1016/j.engappai.2023.107063
87. Xue Q, Li G, Zhang YJ, Shen SQ, Chen Z, Liu YG (2021) Fault diagnosis and abnormality detection of lithium-ion battery packs based on statistical distribution. J Power Sources 482:228964. https://doi.org/10.1016/j.jpowsour.2020.228964
88. Yan GX, Chen J, Bai Y, Yu CQ, Yu CM (2022) A survey on fault diagnosis approaches for rolling bearings of railway vehicles. Processes 10(4):724. https://doi.org/10.3390/pr10040724
89. Yang K, Zhang G, Wang YW, Cai S Finite element analysis on contact stress of high-speed railway bearings. IOP Conf Ser: Mater Sci Eng 1:012073
90. Yang XY, Zheng Y, Zhang Y, Wong DSH, Yang WD (2022) Bearing remaining useful life prediction based on regression shapalet and graph neural network. IEEE Trans Instrum Meas 71:3505712. https://doi.org/10.1109/tim.2022.3151169
91. Yang ZS, Wang YH, Kong CZ (2021) Remaining useful life prediction of lithium-ion batteries based on a mixture of ensemble empirical mode decomposition and GWO-SVR model. IEEE Trans Instrum Meas 70:2517011. https://doi.org/10.1109/tim.2021.3125108
92. Yi C, Li YQ, Huo XM, Tsui KL (2022) A promising new tool for fault diagnosis of railway wheelset bearings: SSO-based Kurtogram. ISA Trans 128:498–512. https://doi.org/10.1016/j.isatra.2021.09.009
93. Yu B, Wang L (2024) Online short-term aging status prediction of SiC MOSFETs for DC solid-state power controller using adaptive variable time-steps metabolic gray model. IEEE Trans Power Electron 39(10):12456–12469. https://doi.org/10.1109/tpel.2024.3414323
94. Yu HJ, Dai HL, Tian GD, Xie YH, Wu BB, Zhu Y, Li HL, Wu H (2020) Big-data-based power battery recycling for new energy vehicles: information sharing platform and intelligent transportation optimization. IEEE Access 8:99605–99623. https://doi.org/10.1109/access.2020.2998178
95. Zang Y, Wei SG, Cai BG, Wang HS, Pecht MG (2019) Methods for fault diagnosis of high-speed railways: a review. Proc Inst Mech Eng Part O-J Risk Reliab 233(5):908–922. https://doi.org/10.1177/1748006x18823932
96. Zeng XQ, Li M, Abd El-Hady D, Alshitari W, Al-Bogami AS, Lu J, Amine K (2019) Commercialization of lithium battery technologies for electric vehicles. Adv Energy Mater 9(27):1900161. https://doi.org/10.1002/aenm.201900161
97. Zhang JL, Hu JB, You HL, Jia RX, Wang XW, Zhang XW (2021) A remaining useful life prediction method of IGBT based on online status data. Microelectron Reliab 121:114124. https://doi.org/10.1016/j.microrel.2021.114124

98. Zhang KL, Huang W, Hou XY, Xu JH, Su RD, Xu HY (2021) A fault diagnosis and visualization method for high-speed train based on edge and cloud collaboration. Appl Sci-Basel 11(3):1251. https://doi.org/10.3390/app11031251
99. Zhang WW, Jia JF, Pang XQ, Wen J, Shi YH, Zeng JC (2024) An improved transformer model for remaining useful life prediction of lithium-ion batteries under random charging and discharging. Electronics 13(8):1423. https://doi.org/10.3390/electronics13081423
100. Zhang ZH, Ding JM, Lin JH (2019) Train wheelset bearing multifault impulsive component separation using hierarchical shift-invariant dictionary learning. Shock Vib 2019:5697137. https://doi.org/10.1155/2019/5697137
101. Zhao Y, Liu P, Wang ZP, Zhang L, Hong JC (2017) Fault and defect diagnosis of battery for electric vehicles based on big data analysis methods. Appl Energy 207:354–362. https://doi.org/10.1016/j.apenergy.2017.05.139
102. Zhou XW, Zhang H, Hao X, Liao X, Han QK (2019) Investigation on thermal behavior and temperature distribution of bearing inner and outer rings. Tribol Int 130:289–298. https://doi.org/10.1016/j.triboint.2018.09.031
103. Zhou YP, Huang MH (2016) Lithium-ion batteries remaining useful life prediction based on a mixture of empirical mode decomposition and ARIMA model. Microelectron Reliab 65:265–273. https://doi.org/10.1016/j.microrel.2016.07.151
104. Zhou YT, Wang YN, Wang K, Kang L, Peng F, Wang LC, Pang JB (2020) Hybrid genetic algorithm method for efficient and robust evaluation of remaining useful life of supercapacitors. Appl Energy 260:114169. https://doi.org/10.1016/j.apenergy.2019.114169
105. Zhu WC, Ni GX, Cao YP, Wang H (2021) Research on a rolling bearing health monitoring algorithm oriented to industrial big data. Measurement 185:110044. https://doi.org/10.1016/j.measurement.2021.110044

Chapter 2
Feature Extraction of Bearing Vibration Signal

2.1 Introduction

Bearings are crucial components within electromechanical systems, particularly in intelligent transport systems such as autonomous vehicles, high-speed trains, and drones [12, 15, 24]. These applications demand high performance from bearings, as long periods of continuous, high-speed operation can lead to failures due to fatigue wear [3, 18]. This context underscores the importance of feature extraction from bearing vibration signals for the development of PHM in IES. It enables early fault detection and prevents potential system failures.

Due to a range of factors including mechanical load variations, inherent bearing wear, operational conditions, and external environmental impacts, bearing vibration signals are characterized by non-stationarity, non-linearity, and randomness [1, 11]. These complex characteristics make it challenging to effectively identify bearing faults using original acoustic signals alone. Consequently, it is crucial to extract sensitive state features that are critical for the accurate identification of bearing faults [10]. Over recent decades, researchers have extensively utilized vibration signal analysis techniques for condition monitoring of bearings. To decode the temporal frequency information of bearing vibration signals, scholars have adopted several frequency-domain analysis methods, such as Short Time Fourier Transformation (STFT) [28], Hilbert-Huang Transform (HHT) [16], WT [29], and Wavelet Packet Transform (WPT) [9]. These methods provide a more nuanced understanding of the dynamic behaviors influencing bearing performance and health.

Most commonly used signal decomposition methods rely on modal analysis, including techniques such as EMD [2, 13], EEMD [31], VMD [8], and CEEMDAN [30]. These methods enable a deeper investigation into the intricate behaviors and anomalies within bearing signals.

Deep learning has emerged as a significant tool in intelligent feature extraction from bearing vibration signals, thanks to its ability to identify features autonomously from raw data using multiple layers of non-linear data-processing units [22]. Ye

H. Liu et al., *Prognostics and Health Management for Intelligent Electromechanical Systems*, https://doi.org/10.1007/978-981-96-7218-9_2

Table 2.1 Different feature extraction methods with advantage and disadvantage

Category	Advantage	Disadvantage
Frequency domain feature extraction	Frequency features are highly sensitive to the detection of early-stage faults, especially when faults change	Frequency analysis may contain a large amount of redundant information, especially in cases of complex signals, which can lead to inefficiencies in data processing
Decomposition-based feature extraction	The ability to automatically deter-mine the modes of decomposition is particularly effective for the analysis of nonlinear signals	Mode mixing issues may occur,, potentially reducing the effectiveness of the decomposition and the accuracy of feature extraction
Deep learning feature extraction	By automatically learn useful features and capture complex time–frequency relationships, it effectively identifying early and subtle fault signals	If there is insufficient data or the model is too complex, deep learning models are prone to overfitting

[26] leveraged a CNN to extract spatial features. Wang [23] introduced a novel network that combines CNN and LSTM to fuse spatial and temporal features, thereby improving the signal's feature representation for more effective fault identification. Cui [5] developed an advanced solution using an Improved Stack AE, which demonstrated superior feature extraction capabilities and accelerated network convergence. Furthermore, Yu [27] applied a transfer learning-based method to bearing fault diagnosis, improving the quality of features while decreasing reliance on labeled data.

Table 2.1 provides a comprehensive summary of the advantages and disadvantages of three principal feature extraction methods for bearing vibration signals. Subsequent sections of this chapter delve into a detailed examination of these common methods: STFT, EMD, and CNN. The evaluation process includes analyzing the extracted features using designated evaluation indices and assessing the classification performance of each feature type by integrating them with real-world data.

2.2 Data Acquisition

In this chapter, experiments on bearing vibration feature extraction are conducted using the high-frequency CWRU dataset [19]. Figure 2.1 illustrates the setup of the CWRU bearing fault test rig, and detailed information about these bearing faults is provided in Table 2.2.

Bearing faults in rotating machinery typically exhibit periodic characteristics. The equation used to determine the number of sampling points is as follows:

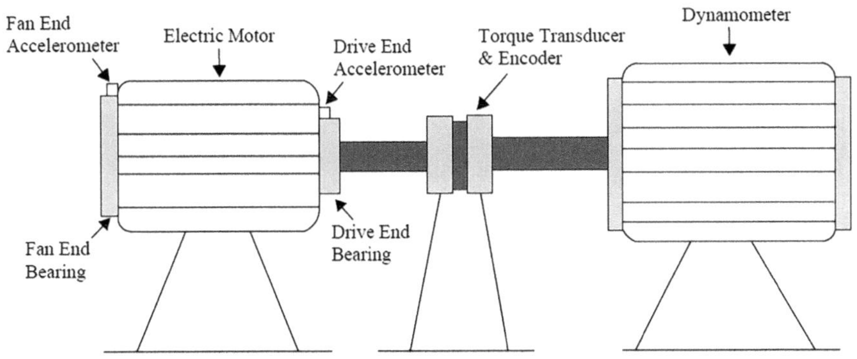

Fig. 2.1 Rolling bearing failure test

Table 2.2 CWRU experiment sample description

Bearing condition	Fault diameter (inch)	Label
Rolling element fault	0.007	1
Rolling element fault	0.014	2
Rolling element fault	0.021	3
Inner race fault	0.007	4
Inner race fault	0.014	5
Inner race fault	0.021	6
Outer race fault	0.007	7
Outer race fault	0.014	8
Outer race fault	0.021	9
Normal	–	10

$$N = \frac{f \times 60}{w} \tag{2.1}$$

where N indicates the total sampling points, f represents the sampling frequency, and w refers to the rotational speed.

Data augmentation is a widely employed technique in data processing that enhances the size and diversity of datasets by applying a series of transformations to original data. This practice improves the generalization ability and robustness of deep learning models. For the one-dimensional time-series data addressed in this study, an overlapping sampling method is utilized for data augmentation. This method involves setting the width of a sliding window equal to the length of a sample, treating the sequence within this window as a single sample. The window then shifts right across the entire dataset by a step size D, selecting the next sample. The formula for calculating the number of samples is as follows:

$$N = \frac{L - l}{D} + 1 \tag{2.2}$$

Fig. 2.2 The process of data enhancement

where L signifies the overall length of the sequence, D denotes the step size, and N represents the total number of samples. Additionally, l refers to the length of an individual sample, as outlined in Eq. (2.3):

$$l = \frac{f}{f_{cycle}} \tag{2.3}$$

The process of overlapping sampling is illustrated in Fig. 2.2. In this scenario, the rotational frequency of the bearing is about 29.95 Hz. the length of the bearing should be at least 400. Utilizing data augmentation techniques, the samples are divided using a method of repetitive overlaying. With L approximately equal to 120,000, l set at 2024, and D set at 1191, N for one type of fault is 100.

The validation utilizes rolling element fault data collected at motor speeds of 1797 rpm under a motor load of 0 hp. Fault diameters of 0.007, 0.014, and 0.021 inches are examined, with data sampled at 12 kHz. In particular, the original time-domain waveform diagrams for four types of rolling bearing signals are shown in Fig. 2.3.

2.3 Frequency Domain Feature Extraction

2.3.1 The Theoretical Basis of Continuous Wavelet Transform

The Continuous Wavelet Transform (CWT) overcomes the fixed window size limitation of the STFT by employing a "time–frequency" window that dynamically adapts to different frequencies [17, 25]. The wavelet basis function, $\varphi_{a,b}(t)$, consists of wavelet basis functions $\varphi(t)$ (also known as mother wavelets), which generate scaled and translated versions (referred to as daughter wavelets), as depicted in Eq. (2.4).

$$\varphi_{a,b}(t) = \frac{1}{\sqrt{a}}\varphi\left(\frac{t-b}{a}\right) \tag{2.4}$$

Fig. 2.3 Time-domain waveform diagrams for four types of rolling bearing signals

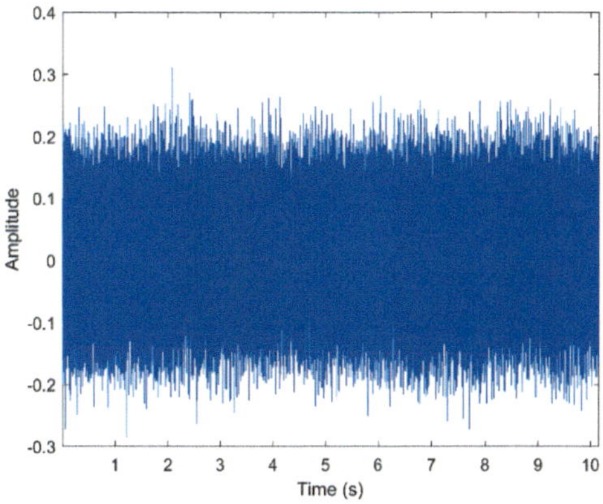

(a) Normal

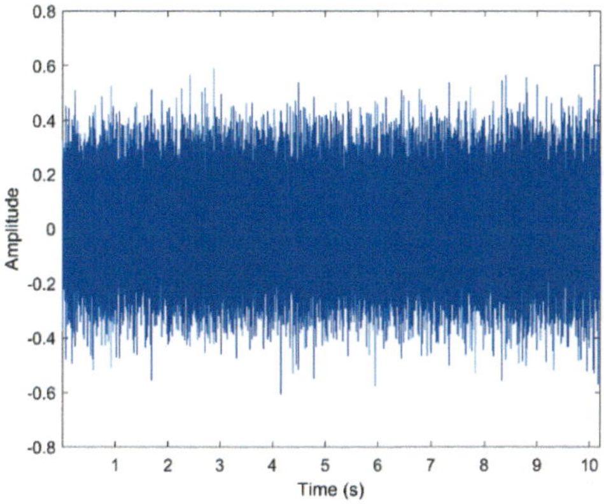

(b) Rolling element fault measuring 0.007 ich

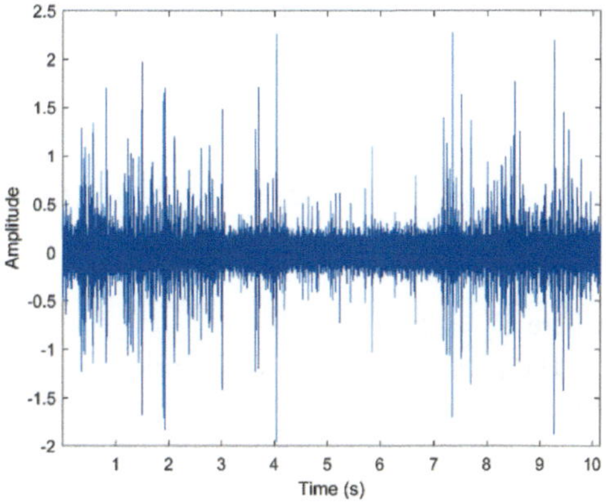

(c) Rolling element fault measuring 0.014 ich

Fig. 2.3 (continued)

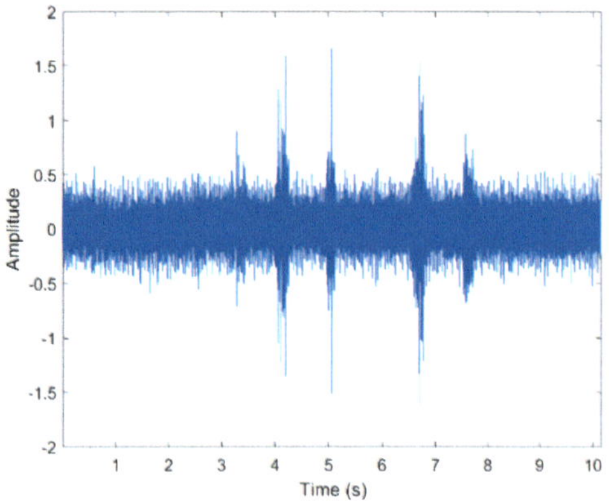

(d) Rolling element fault measuring 0.021 ich

where a > 0 is the scale parameter that determines the scaling characteristics of the wavelet, and b determines the wavelet's position by controlling its shift, $a^{-1/2}$ employed to ensure the conservation of energy in transformation.

$$\left\|\varphi_{a,b}(w)\right\|^2 = \int_{-\infty}^{\infty} \left|\varphi_{a,b}(t)\right|^2 dt = \int_{-\infty}^{\infty} |\varphi(t)|^2 dt \tag{2.5}$$

For a square-integrable signal $f(t)$ within L^2(R), the CWT of $f(t)$ is performed as follows:

$$CWT_f(a, b) = \frac{1}{\sqrt{a}} \int f(t)\varphi'\left(\frac{t-b}{a}\right)dt = \int f(t)\varphi'_{a,b}(t)dt = \left\langle f(t), \varphi_{a,b}(t)\right\rangle \tag{2.6}$$

Upon $CWT_f(a, b)$ executing the inverse transform, it can be established that:

$$x(t) = \frac{1}{c\varphi} \int_{-\infty}^{\infty} \varphi_{a,b}(t) CWT_x(a, b) \frac{1}{a^2} dadb \tag{2.7}$$

where $C_\varphi = 2\pi \int_{-\infty}^{\infty} \frac{|\varphi(w)|}{|w|} dw$.

The process begins with the selection of an appropriate wavelet basis function. Afterward, a fixed scale factor is chosen to compare with the initial segment of the signal. Following this, the wavelet coefficients are calculated using the CWT formula. The next step involves adjusting the translation factor to shift the wavelet along the time axis, repeating the previous calculations. This completes the first analysis cycle. The scale factor is then increased, and the steps are repeated for a second cycle. This iterative process continues, adjusting the scale and translation parameters as needed until the analysis meets the desired experimental criteria.

2.3.2 Feature Extraction

When applying the CWT, a larger scaling factor in the wavelet basis function is suited for extracting low-frequency features from the signal. Conversely, a smaller scaling factor is ideal for high-frequency feature extraction. Choosing an appropriate wavelet basis function is crucial for ensuring the effectiveness and efficiency of the wavelet transform, especially in the analysis of bearing fault signals. The Cmor wavelet, a complex variant of the Morlet wavelet, was selected for its excellent resolution and adaptability in both time and frequency domains [7], making it suitable for capturing the time–frequency features of bearing signals.

A sample of size 2048 was randomly selected, covering four states: normal, 0.007, 0.014 and 0.021-inch fault in the rolling elements. The time–frequency waveform and CWT feature extraction results for each state are shown in Figs. 2.4, 2.5, 2.6 and 2.7. The characteristics of the signals and the corresponding CWT results for each state are analyzed as follows:

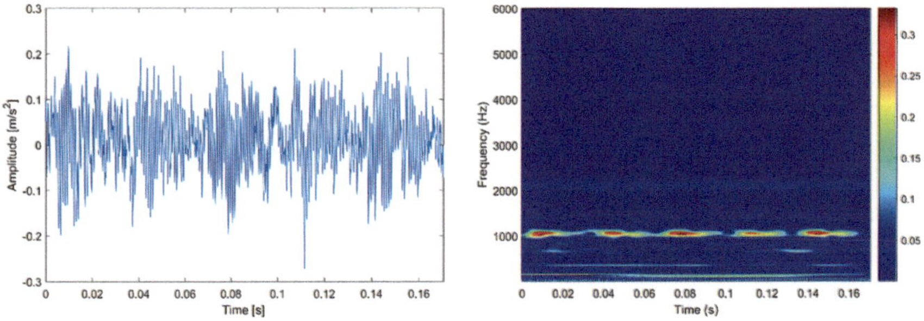

(a) signal waveform (b) CWT feature extraction results

Fig. 2.4 Normal

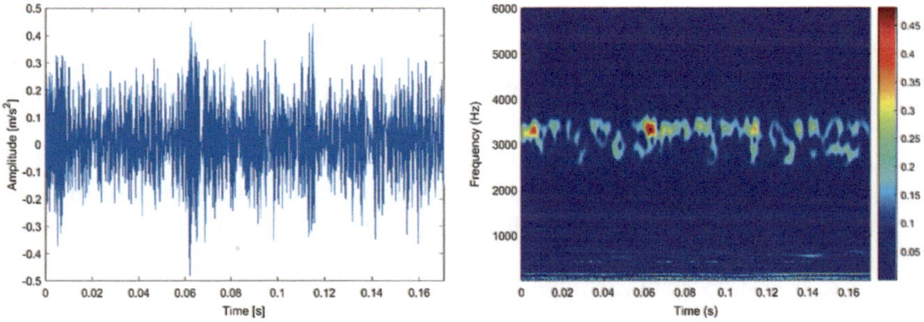

(a) Signal waveform (b) CWT feature extraction results

Fig. 2.5 Rolling element fault with 0.007 in

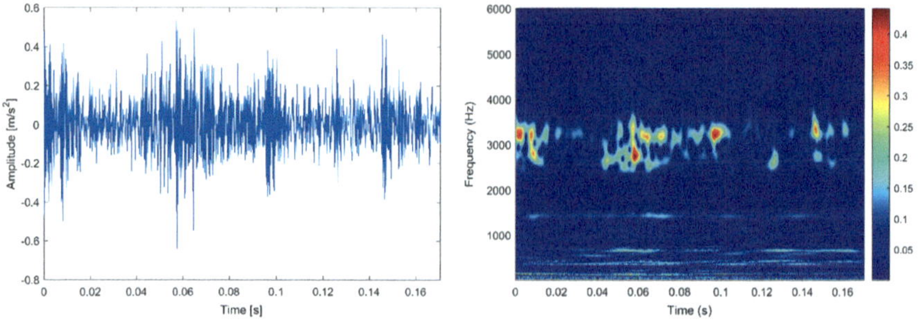

(a) Signal waveform (b) CWT feature extraction results

Fig. 2.6 Rolling element fault with 0.014 in

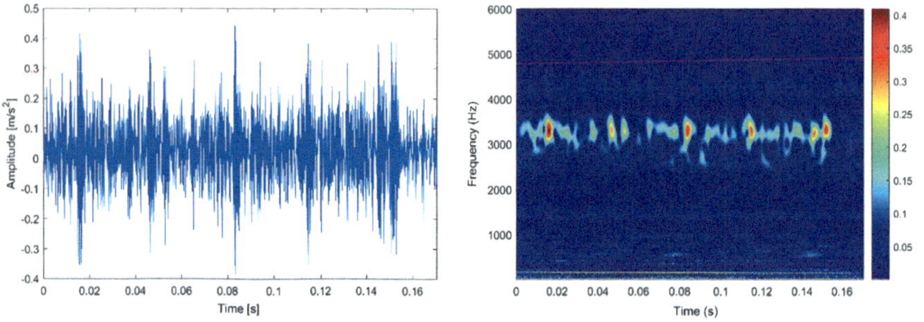

(a) Signal waveform (b) CWT feature extraction results

Fig. 2.7 Rolling element fault with 0.021 in

(1) Figures 2.4a, 2.5a, 2.6a and 2.7a illustrate the waveforms of the vibration signals, revealing distinct patterns for each bearing condition. In the normal state, the vibration pattern appears relatively calm, with low amplitude fluctuations mainly ranging between − 0.1 and + 0.1 m/s^2, and no prominent spikes or abnormal vibrations. When a 0.007-inch rolling element fault is present, the amplitude increases, showing more fluctuations and sharp peaks, which are typical indicators of localized contact issues caused by the fault. As the fault diameter increases to 0.014 inches, the amplitude grows further, with more pronounced abnormal fluctuations compared to the normal state. At a fault size of 0.021 inches, the signal exhibits the highest vibration amplitude, with fluctuations ranging from − 0.5 to + 0.5 m/s^2 and frequent, sharp, high-magnitude peaks.
(2) Figures 2.4b, 2.5b, 2.6b and 2.7b display the CWT signal processing results, where fault characteristic frequencies and their harmonics can be identified. In the CWT spectrum of the normal bearing, the energy distribution is uniform and concentrated in the lower frequency region, indicating the absence of

significant transients or fault-related frequency components. For the 0.007-inch rolling element fault, some high-energy regions emerge, primarily concentrated between 1000 and 2000 Hz. These high-energy zones suggest that the fault caused an increase in energy at specific frequencies. With the 0.014-inch rolling element fault, the energy concentration spreads across a wider frequency band, especially between 2000 and 5000 Hz, indicating that the fault's impact is more extensive and may involve additional dynamic components. The 0.021-inch rolling element fault exhibits very prominent and more frequent high-energy regions, particularly in the high-frequency range of 2500–5000 Hz. These frequent high-energy areas reflect the broader frequency distribution of vibrations caused by the fault.

By comparing the CWT spectrograms of the normal bearing with those of varying fault sizes, it is evident that as the fault size increases, the fault characteristic frequency bands become wider, and the high-energy regions become more frequent and pronounced. These changes highlight the significant impact of faults on the overall dynamic behavior of the bearing, with larger faults exerting a broader influence. The ability of CWT to capture these features effectively reflects the relationship between fault characteristic frequencies and time, making it a valuable tool for quantitative assessment in practical applications. This capability facilitates timely maintenance and replacement, helping to prevent more severe electromechanical system failures and downtime.

2.3.3 Feature Evaluation

To evaluate the effectiveness of CWT in capturing features for bearing fault diagnosis, this section employs three quantitative and qualitative metrics: Mean Squared Error (MSE), Signal-to-Noise Ratio (SNR) [6], and Spectral Entropy (SpecEn) [4]. The corresponding calculation formulas are provided in Eqs. (2.8)–(2.10).

$$MSE = \sum_{l=1}^{L} [x(l) - \hat{x}(l)]^2 / L \tag{2.8}$$

$$SNR = 10 \lg \frac{\sum_{l=1}^{L} x^2(l)}{\sum_{l=1}^{L} [x(l) - \hat{x}(l)]^2} \tag{2.9}$$

$$SpecEn = -\sum_{i=1}^{N} p_i * \log p_i \tag{2.10}$$

where $x(l)$ denotes the raw vibration signal, while $x(l)$ refers to the reconstructed vibration signal. L represents the signal length. Furthermore, p_i is defined as the ratio of the ith power spectrum S_i to the total power spectrum, where $i = 1, 2, \ldots, N$, and N corresponds to the total number of samples.

The results of MSE, SNR, and SpecEn calculations after CWT processing for the four conditions: normal, 0.007, 0.014, and 0.021-inch fault, are presented in Table 2.3. The findings from Table 2.3 reveal that:

(1) The MSE values between the original vibration signals and the signals reconstructed from the CWT coefficients for all four conditions are less than 0.015. This low reconstruction error indicates that the CWT features effectively captured the essential information from the original signals.
(2) The SNR values for all four conditions are greater than 0, indicating that the reconstructed signals contain more useful information relative to noise after CWT feature extraction. This means that the fault characteristics are more pronounced in the signal, making them easier to detect.
(3) The SpecEn values gradually rise as the rolling element transitions from a normal state to the maximum fault size, indicating that the spectral distribution of the signal becomes more uniform and complex as the fault severity increases. This suggests that as the fault develops, the signal's frequency distribution becomes broader and more random.

The above quantitative analysis further demonstrates the high sensitivity and accuracy of the time–frequency feature extraction method, CWT, which is well-suited for early fault detection and classification. This capability aids maintenance teams in performing predictive maintenance, optimizing maintenance schedules, reducing unnecessary downtime, and significantly enhancing equipment reliability and efficiency.

Table 2.3 Feature evaluation results

Rolling element condition (inch)	MSE	SNR	SpecEn
0	0.01369	0.4069	4.5597
0.007	0.01466	0.9081	5.8871
0.014	0.01280	0.8706	6.0733
0.021	0.01990	0.8715	6.0296

2.4 Decomposition-Based Feature Extraction

2.4.1 The Theoretical Basis of Variational Modal Decomposition

VMD is an adaptive approach that separates a signal into several Intrinsic Mode Functions (IMFs), each corresponding to a specific central frequency [14, 20]. In VMD, each IMFs is considered a signal with a finite bandwidth. The computation process is carried out through the following steps:

Step 1: Define the variational problem, where the $x(t)$ is decomposed into K mode components $u_k(t)$ using VMD, expressed as:

$$u_k(t) = A_k(t) \cos \varphi_k(t) \tag{2.11}$$

where $A_k(t)$ represents the amplitude of the signal component. t denotes time. φ_k is the phase. $w_k(t)$ is the instantaneous frequency, calculated by $w_k(t) = d\varphi_k(t)/dt$.

For each IMF, the Hilbert transform is used to derive the single-sided spectrum, which is shifted to the base frequency band through exponential correction. Gaussian smoothing is then applied to demodulate the signal and estimate the bandwidth of each mode. The resulting function is expressed as:

$$\begin{cases} \min\limits_{\{u_k\}\times\{w_k\}} \left\{ \sum\limits_{k=1}^{K} \left\| \partial_t \left[\delta(t) + \frac{j}{\pi t} * u_k(t) \right] e^{-jw_k t} \right\|_2^2 \right\} \\ \sum\limits_{k=1}^{K} u_k = x(t) \end{cases} \tag{2.12}$$

where $\delta(t)$ represents the Impulse function and $\{w_k\} = \{w_1, w_2, \ldots, w_K\}$ is the set of central frequencies.

Step 2: The augmented Lagrangian function is formulated by incorporating the Lagrange multiplier $\lambda(t)$ and the penalty term α, leading to the following expression:

$$\begin{aligned} L(\{u_k\}, \{w_k\}, \lambda) = \alpha \sum_{k=1}^{K} \left\| \partial_t \left[\delta(t) + \frac{j}{\pi t} * u_k(t) \right] e^{-jw_k t} \right\| \\ + \left\| x(t) - \sum_{k=1}^{K} u_k \right\| + \left\langle \lambda(t), x(t) - \sum_{k=1}^{K} u_k \right\rangle \end{aligned} \tag{2.13}$$

Step 3: Utilizing the Alternating Direction Method of Multipliers to solve Eq. (2.13) yields the following formulas:

$$\hat{u}_k^{n+1}(w) = \frac{\hat{x}(w) - \sum_{i=1, i \neq k}^{K} \hat{u}_i(w) + \frac{\hat{\lambda}(w)}{2}}{1 + 2\alpha(w - w_k)^2} \tag{2.14}$$

$$w_k^{n+1} = \frac{\int_0^\infty w |\hat{u}_k(w)|^2 dw}{\int_0^\infty |\hat{u}_k(w)|^2 dw} \tag{2.15}$$

$$\hat{\lambda}^{n+1}(w) = \hat{\lambda}^n(w) + \tau \left(\hat{x}(w) - \sum_{k=1}^{K} \hat{u}_k^{n+1}(w) \right) \tag{2.16}$$

where $\hat{u}_k^{n+1}(w)$, $\hat{x}(w)$, and $\hat{\lambda}(w)$ represent the Fourier transforms. τ denotes the noise component.

Step 4: To obtain the optimal solution for Eq. (2.12), the Alternating Direction Method of Multipliers is applied to iteratively update $\{u_k\}$, $\{w_k\}$, and λ. This iterative process ultimately achieves the decomposition of the original signal.

2.4.2 Feature Extraction

Randomly selected vibration signal samples for normal, 0.007, 0.014, and 0.021-inch rolling element faults were processed using VMD, resulting in the corresponding IMF components. The time-domain waveforms and spectrograms of these IMFs are displayed in Figs. 2.8, 2.9, 2.10 and 2.11. The analysis of the IMF time-domain waveforms and spectrograms for each condition is as follows:

(1) From Figs. 2.8a, 2.9a, 2.10a and 2.11a, it is evident that under normal conditions, the waveforms appear relatively smooth and regular, indicative of typical operating conditions. With the progression of fault severity, the waveforms exhibit more irregular variations and higher amplitudes. This progression indicates the

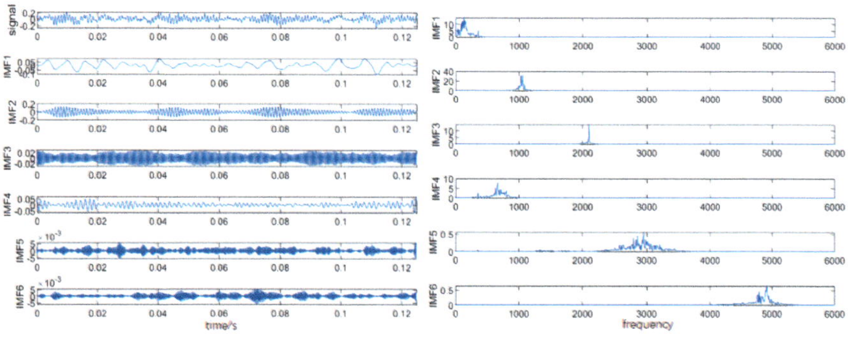

(a) Time domain diagram (b) Spectrum diagram

Fig. 2.8 Normal

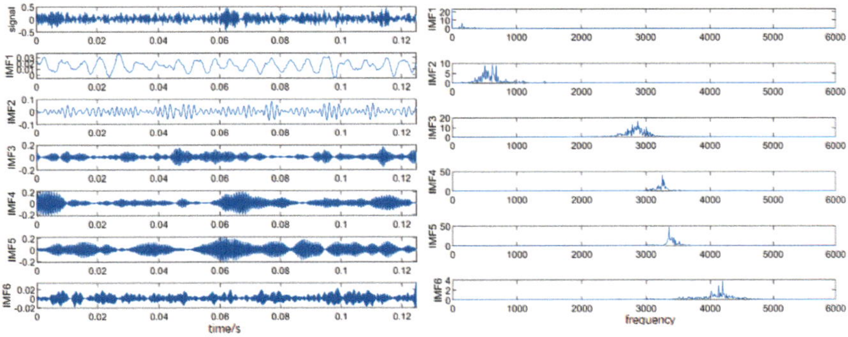

(a) Time domain diagram (b) Spectrum diagram

Fig. 2.9 Rolling element fault with 0.007 in

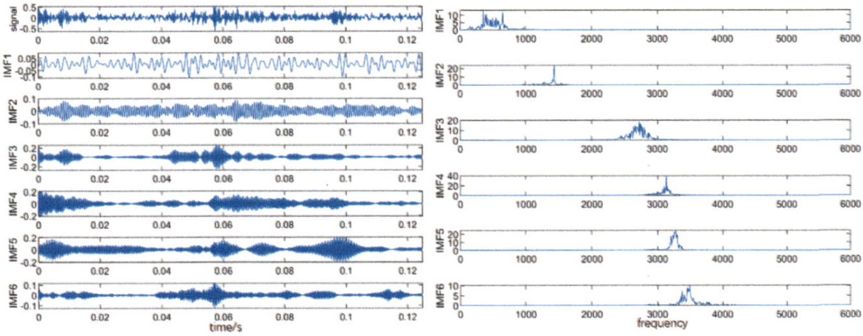

(a) Time domain diagram (b) Spectrum diagram

Fig. 2.10 Rolling element fault with 0.014 in

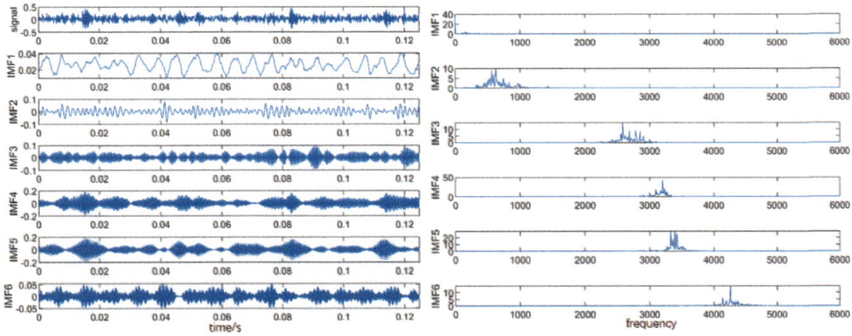

(a) Time domain diagram (b) Spectrum diagram

Fig. 2.11 Rolling element fault with 0.021 in

deterioration of the bearing's rolling elements, where larger faults disrupt the regularity and smoothness of operation.

(2) From Figs. 2.8b, 2.9b, 2.10b and 2.11b, it can be observed that the spectra typically exhibit fewer prominent peaks with relatively small magnitudes under normal conditions, indicating minimal vibration caused by defects and no clear fault frequencies. The rolling element faults may not introduce sufficient frequency components in the vibration signal to show significant features in IMF1. Such faults are more pronounced in lower-frequency IMFs. In the spectrum of the 0.007-inch rolling element fault signal, distinct peaks are visible, particularly in IMF2, IMF3, and IMF5, pointing to specific fault frequencies. The 0.014-inch fault signal spectrum shows even more prominent peaks, especially in IMF2, IMF3, IMF4, and IMF5, indicating increased fault severity. For the 0.021-inch fault, multiple IMFs, particularly IMF2, IMF3, IMF4, and IMF5, exhibit significant peaks, confirming the severity of the fault.

The application of VMD technology for the multi-scale decomposition of vibration signals effectively extracts useful fault feature information from complex signals, providing a powerful tool. The variations in each IMF component reveal different stages and characteristics of fault development, aiding in more precise health condition assessment and fault prediction.

2.4.3 *Feature Evaluation*

To assess the performance of VMD in extracting features for diagnosing bearing faults, the analysis incorporates kurtosis (Kur), energy entropy (EEn), and sample entropy (SampEn) as metrics to describe the operational state of the bearings. These metrics are applied to the VMD-decomposed bearing vibration signals to assess their ability to capture fault characteristics accurately.

(1) The Kur reflects the distribution characteristics of a vibration signal and describes the peakedness of its waveform. The mathematical expression for kurtosis is:

$$Kur = \frac{E(x-\mu)^4}{\sigma^4} \tag{2.17}$$

where x represents the analyzed vibration signal, μ and σ is the mean and standard deviation, respectively.

(2) The EEn quantifies the regularity of a time series through the IMFs. The definition of energy entropy is as follows:

$$EEn = -\sum_{i=1}^{n} p_i \ln p_i \tag{2.18}$$

where $p_i = E_i/E$ is the normalized energy. Here, $E_i = \sum_{j=1}^{m} |c_{ij}|^2$ denotes the energy of each IMF component and $E = \sum_{i=1}^{n} E_i$ is the total energy.

(3) The SampEn quantifies the complexity of a time series by assessing the likelihood of new patterns emerging within the signal. The computation process is outlined as follows:

$$SampEn(m, r, L) = -\ln \frac{\Phi_{m+1}(r)}{\Phi_m(r)} \tag{2.19}$$

where m is the embedding dimension independent of the time series, r is the embedding dimension independent of the time series. Φ_m represents the average count of similar patterns, which is given by:

$$\Phi_m(r) = \frac{1}{L-m+1} \sum_{i=1}^{L-m+1} \left[\frac{1}{L-m} num\{d[x(i), x(j)] < r\} \right] \tag{2.20}$$

where $num\{d[x(i), x(j)] < r\}$ denotes the count of pairs where the distance. $i \neq j$ indicates that self-matching is excluded. In this method, it is recommended to set $m = 2$ and $r = (0.1–0.25)$SD, where SD refers to the standard deviation.

The Kur, EEn, and SampEn results for VMD decomposition under the four conditions of normal, rolling element faults of 0.007, 0.014 and 0.021 inch are presented in Table 2.4. As seen in Table 2.4, the following observations can be made:

(1) Higher Kur values signify the presence of more impulsive components in the signal. In the normal state, the Kur values for all IMF components are relatively low, indicating a smooth signal with minimal spikes. As the severity increases,

Table 2.4 Evaluation of VMD decomposition features for bearings

Metrics	Condition (inch)	IMF1	IMF2	IMF3	IMF4	IMF5	IMF6
Kur	0	3.2086	2.4178	1.7515	3.7947	3.0123	3.6160
	0.007	2.3684	3.0345	3.5302	4.2298	2.6481	3.0476
	0.014	2.9951	2.1309	5.8324	3.8336	3.6711	3.1018
	0.021	3.1437	3.1524	6.8141	4.5341	3.4361	3.3879
EEn	0	0.3545	0.2951	0.1456	0.1600	0.0041	0.0039
	0.007	0.0872	0.1451	0.2898	0.3668	0.3631	0.0233
	0.014	0.2190	0.1923	0.3434	0.3461	0.3601	0.1480
	0.021	0.1902	0.1700	0.2699	0.2593	0.1272	0.0386
SampEn	0	0.3053	0.5461	0.26007	0.5967	0.7961	0.6149
	0.007	0.4351	0.6010	0.3405	0.1733	0.1913	0.7193
	0.014	0.5848	0.4674	0.3994	0.1847	0.1376	0.2310
	0.021	0.4638	0.6191	0.4619	0.1991	0.2129	0.4427

Kur values rise significantly, especially for IMF3 and IMF4. This increase highlights the development of sharp peaks associated with the fault. These results suggest that IMF3 and IMF4 are particularly effective for detecting rolling bearing faults.

(2) Higher EEn values indicate a more uniform distribution of energy. In the normal state, the EEn values for all IMFs are relatively high, reflecting a uniform energy distribution typical of a well-functioning bearing. When a 0.007-inch fault occurs, EEn values decrease for all IMFs except IMF1, suggesting that energy starts to concentrate in specific frequency components, likely due to initial bearing damage. With a 0.014-inch fault, most IMFs show a further decrease in EEn, particularly IMF4, and IMF5, indicating that the fault has intensified, leading to more concentrated energy. In the case of a 0.021-inch fault, EEn values are generally lower than those at 0.014 inches, especially in IMF5 and IMF6, which reflects a more pronounced concentration of energy and increased system instability.

(3) Higher SampEn demonstrates greater signal complexity. In the normal state, the SampEn values for all IMFs show a certain level of complexity, with IMF2, IMF4, and IMF5 displaying particularly high values. This suggests that the bearing signals in the normal condition include multiple frequency components and are relatively complex. As the severity increases, the SampEn values for most IMFs rise. is due to the inclusion of new frequency components and irregularities caused by the fault, indicating that as the fault worsens, the complexity and unpredictability of the system also increase.

In conclusion, the IMF components obtained from VMD decomposition can be utilized to compute various statistical metrics, such as Kur, EEn, and SampEn. These indicators aid in evaluating the health status of bearings from different perspectives. VMD offers a robust approach for analyzing and interpreting complex signals in electromechanical systems, significantly enhancing the capability for fault detection. This, in turn, provides strong support for preventive maintenance and fault diagnosis.

2.5 Deep Learning Feature Extraction

2.5.1 The Theoretical Basis of Convolutional Neural Network

The CNN performs convolutional operations [21, 32]. In these networks, convolutional and pooling layers are typically alternated to extract feature data. Activation layers are usually positioned after the convolutional and fully connected layers to enhance the network's ability to perform nonlinear mappings. The fully connected layers then map these extracted features nonlinearly and output them to classifiers, which commonly employ the SoftMax function. A schematic of CNN is shown in Fig. 2.12.

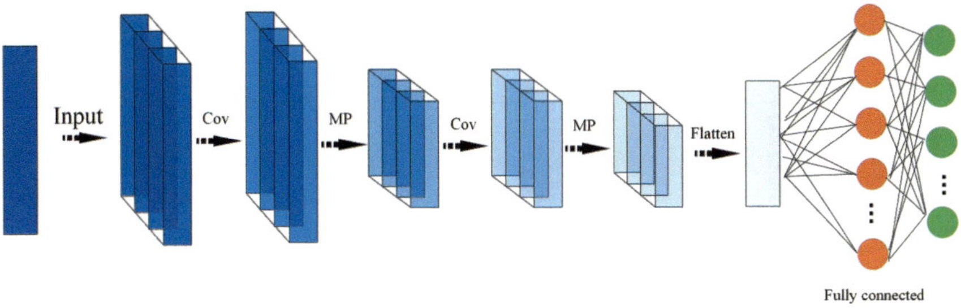

Fig. 2.12 Structure of the CNN

The convolution operation is primarily aimed at feature learning. The convolutional kernel embodies two critical concepts: sparse connectivity and weight sharing. This kernel acts as a trainable feature extractor, with its output commonly referred to as a feature map. It applies convolution operations to the outputs from the previous layer, passing the extracted features to the subsequent layer. This process facilitates sparse connections and parameter sharing as data is transmitted across layers. The computation in the convolutional layer can be described as follows:

$$x_j^l = f\left(\sum_{i\in M_j}^{n} x_i^{l-1} * k_{i,j}^l + b_j^l\right) \tag{2.21}$$

where x^l, k^l, b^l represents the output, convolution kernel and bias term of layer l, respectively. f is activation function, * denotes convolution operation, M_j represents selected set of input features.

2.5.2 Feature Extraction

Using vibration signal samples from normal conditions and rolling element faults at 0.007, 0.014, and 0.021 inch, divided into four categories. The dataset is split evenly into training and test sets with a 1:1 ration. The convolutional layers of the CNN have kernel sizes of 8 and 16. The grayscale feature maps extracted from the last convolutional layer are displayed in Fig. 2.13. These feature maps encapsulate the key information captured by the CNN when processing bearing vibration or image data.

Figure 2.13 illustrates that the feature map for the normal state, as shown in Fig. 2.13a, displays a uniform or regular pattern, reflecting the consistency of the bearing's vibration or sound signals in the absence of any clear faults. The feature map for the 0.007-inch fault, as shown in Fig. 2.13c, reveals slight abnormal patterns, such as enhanced brightness in certain areas, which may indicate minor changes caused by the fault. The feature maps for the 0.014 and 0.021-inch faults, depicted in Fig. 2.13e,

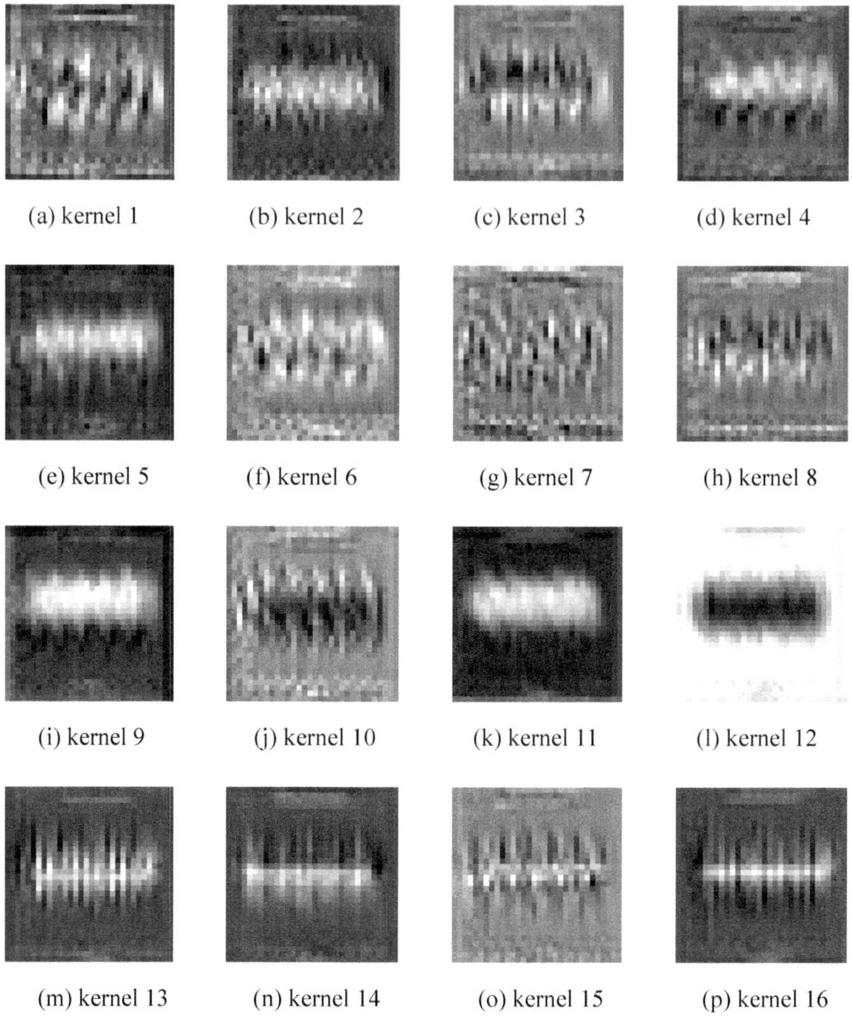

(a) kernel 1 (b) kernel 2 (c) kernel 3 (d) kernel 4

(e) kernel 5 (f) kernel 6 (g) kernel 7 (h) kernel 8

(i) kernel 9 (j) kernel 10 (k) kernel 11 (l) kernel 12

(m) kernel 13 (n) kernel 14 (o) kernel 15 (p) kernel 16

Fig. 2.13 Feature map

exhibit more pronounced and widespread changes. These maps might show brighter areas or chaotic patterns, indicating more severe mechanical damage affecting the bearing's performance.

In conclusion, the analysis demonstrates that the feature maps generated by the CNN are instrumental in the early identification of bearing faults, even when external symptoms are not readily apparent. This is crucial for preventive maintenance and avoiding costly failures in electromechanical systems.

2.5.3 Feature Evaluation

The Softmax classifier is employed in CNN as output layer, which presents the classification results in the form of probabilities, facilitating the final category prediction. The higher the probability value for a particular category, the stronger the inclination towards that category. Fault categories 1, 2, 3, and 4 correspond to normal bearing conditions, and rolling element faults of 0.007, 0.014, and 0.021 inch, respectively. Figure 2.14 illustrates the changes in feature distribution before and after sample recognition by CNN.

Figure 2.14a shows that before processing by CNN, the distribution of categories is quite broad, and the boundaries between categories are not well defined. This indicates that prior to feature extraction, there is considerable overlap of sample features in the space, making classification challenging. Figure 2.14b reveals that the points representing each category are more concentrated, and the separation between different clusters is more distinct. This demonstrates that the CNN has effectively extracted key features that distinguish between different categories and has achieved better separation in the feature space. It preliminarily reflects the effectiveness of the CNN model in feature extraction and classification tasks.

To further verify the multi-class classification performance of the CNN, this section employs Accuracy, Precision, Recall, and F1 Score as evaluation metrics, as outlined in Eqs. (2.22)–(2.25).

$$Accuracy = \frac{TP + TN}{TP + TN + FP + FN} \tag{2.22}$$

$$P = \frac{TP}{TP + FP} \tag{2.23}$$

$$R = \frac{TP}{TP + FN} \tag{2.24}$$

$$F1 = \frac{2PR}{P + R} \tag{2.25}$$

where *TP* refers to the normal samples correctly identified, *FN* denotes normal samples misclassified as faulty, *FP* represents faulty samples misclassified as normal, and *TN* indicates faulty samples correctly identified as faulty.

In this section, for the multi-class task involving four different states of bearings, combinations of any two types are used to generate confusion matrices. From these matrices, n pairs of precision and recall values $(P_1, R_1)(P_2, R_2)\ldots(P_n, R_n)$ are obtained. This leads to the calculation of macro-average precision (macro_P), macro-average recall (macro_R), and the corresponding macro F1 Score (macro_F1). The specific calculation methods are provided in Eqs. (2.26)–(2.28):

Fig. 2.14 Sample distribution diagram

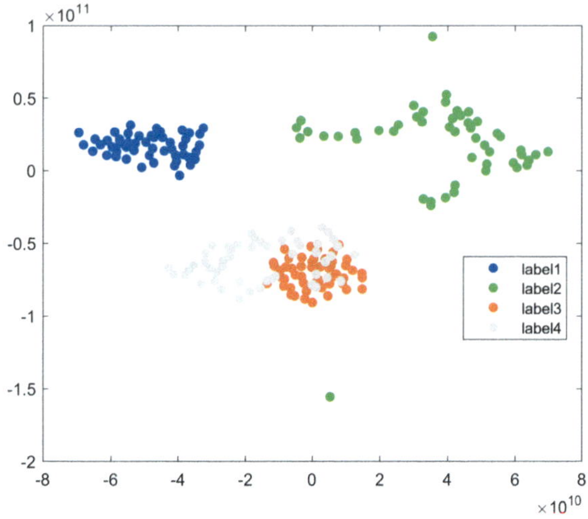

(a) Original sample distribution

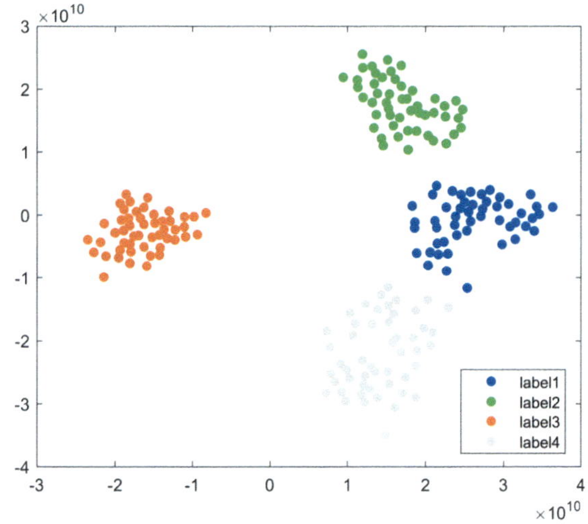

(b) Post-model recognition sample distribution

$$macro\_P = \frac{1}{n}\sum_{i=1}^{n} P_i \tag{2.26}$$

$$macro\_R = \frac{1}{n}\sum_{i=1}^{n} R_i \tag{2.27}$$

$$macro\_F1 = \frac{2 \times macro\_P \times macro\_R}{macro\_P + macro\_R} \tag{2.28}$$

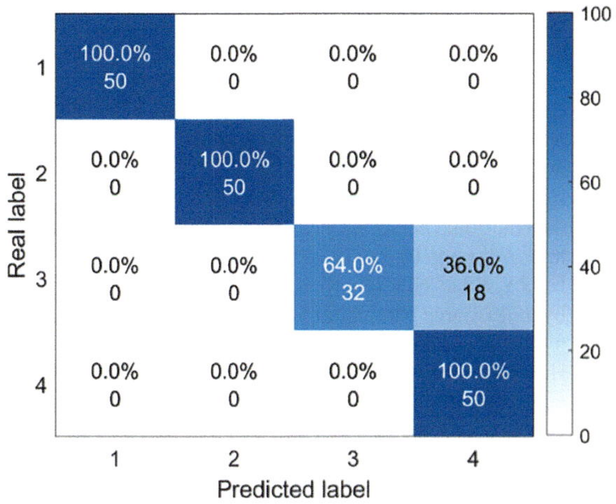

Fig. 2.15 Confusion matrix

The confusion matrix for CNN-based classification prediction of bearing states is displayed in Fig. 2.15, with Acc of 0.9100, macro_P of 0.9300, and macro_F1 of 0.9200.

These results demonstrate that the CNN model exhibits efficient and reliable performance in fault diagnosis tasks, managing to maintain a high precision without sacrificing recall. This makes the model well-suited for applications requiring high accuracy and reliability, such as fault prediction and diagnosis in electromechanical systems.

References

1. Buchaiah S, Shakya P (2022) Bearing fault diagnosis and prognosis using data fusion based feature extraction and feature selection. Measurement 188:110506. https://doi.org/10.1016/j.measurement.2021.110506
2. Chen DY, Lin JH, Li YP (2018) Modified complementary ensemble empirical mode decomposition and intrinsic mode functions evaluation index for high-speed train gearbox fault diagnosis. J Sound Vib 424:192–207. https://doi.org/10.1016/j.jsv.2018.03.018
3. Chen GG, Li Y, Han SY, Shao HD, Yang XK (2023) Intelligent fault diagnosis for high-speed bearing towards unbalanced samples via convolutional weight adaptive network. J Eng Des 1–24. https://doi.org/10.1080/09544828.2023.2261095
4. Civera M, Surace C (2022) An application of instantaneous spectral entropy for the condition monitoring of wind turbines. Appl Sci-Basel 12(3):1059. https://doi.org/10.3390/app12031059
5. Cui ML, Wang YQ, Lin XS, Zhong MY (2021) Fault diagnosis of rolling bearings based on an improved stack autoencoder and support vector machine. IEEE Sens J 21(4):4927–4937. https://doi.org/10.1109/jsen.2020.3030910
6. Du P, Kibbe WA, Lin SM (2006) Improved peak detection in mass spectrum by incorporating continuous wavelet transform-based pattern matching. Bioinformatics 22(17):2059–2065. https://doi.org/10.1093/bioinformatics/btl355

7. Gao J, Song QH, Liu ZQ (2018) Chatter detection and stability region acquisition in thin-walled workpiece milling based on CMWT. Int J Adv Manuf Technol 98(1–4):699–713. https://doi.org/10.1007/s00170-018-2306-1
8. Guo YJ, Yang YD, Jiang SF, Jin XH, Wei YD (2022) Rolling bearing fault diagnosis based on successive variational mode decomposition and the EP Index. Sensors 22(10):3889. https://doi.org/10.3390/s22103889
9. He F, Ye Q (2022) A bearing fault diagnosis method based on wavelet packet transform and convolutional neural network optimized by simulated annealing algorithm. Sensors 22(4):1410. https://doi.org/10.3390/s22041410
10. He ZY, Shao HD, Ding ZY, Jiang HK, Cheng JS (2022) Modified deep autoencoder driven by multisource parameters for fault transfer prognosis of aeroengine. IEEE Trans Ind Electron 69(1):845–855. https://doi.org/10.1109/tie.2021.3050382
11. Hou DM, Qi HY, Luo HL, Wang CP, Yang JT (2022) Comparative study on the use of acoustic emission and vibration analyses for the bearing fault diagnosis of high-speed trains. Struct Health Monit Int J 21(4):1518–1540. https://doi.org/10.1177/14759217211036025
12. Li SS, Zhang YW, Wang XJ, Wang SP, Duan HB (2024) Prescribed-time bearing-based formation control of underactuated ASVs under external disturbance. IEEE Trans Circuits Syst II-Express Briefs 71(3):1196–1200. https://doi.org/10.1109/tcsii.2023.3325435
13. Li Y, Zhou JW, Li HG, Meng G, Bian J (2023) A fast and adaptive empirical mode decomposition method and its application in rolling bearing fault diagnosis. IEEE Sens J 23(1):567–576. https://doi.org/10.1109/jsen.2022.3223980
14. Lian JJ, Liu Z, Wang HJ, Dong XF (2018) Adaptive variational mode decomposition method for signal processing based on mode characteristic. Mech Syst Sig Process 107:53–77. https://doi.org/10.1016/j.ymssp.2018.01.019
15. Liu GH, Hong J, Parker RG (2019) Influence of simultaneous time-varying bearing and tooth mesh stiffness fluctuations on spur gear pair vibration. Nonlinear Dyn 97(2):1403–1424. https://doi.org/10.1007/s11071-019-05056-9
16. Liu ZL, Peng DD, Zuo MJ, Xia JS, Qin Y (2022) Improved Hilbert-Huang transform with soft sifting stopping criterion and its application to fault diagnosis of wheelset bearings. ISA Trans 125:426–444. https://doi.org/10.1016/j.isatra.2021.07.011
17. Prasad A, Maan J, Verma SK (2021) Wavelet transforms associated with the index Whittaker transform. Math Methods Appl Sci 44(13):10734–10752. https://doi.org/10.1002/mma.7440
18. Shi MF, Liu XZ, Wei WB, Liu YB, Liu F, Li GL (2022) Wayside acoustic fault diagnosis of train bearing based on improved linear discriminant analysis. IEEE Sens J 22(14):14228–14236. https://doi.org/10.1109/jsen.2022.3176057
19. Smith WA, Randall RB (2015) Rolling element bearing diagnostics using the Case Western Reserve University data: a benchmark study. Mech Syst Sig Process 64–65:100–131. https://doi.org/10.1016/j.ymssp.2015.04.021
20. Song QY, Jiang XX, Wang S, Guo JF, Huang WG, Zhu ZK (2022) Self-adaptive multivariate variational mode decomposition and its application for bearing fault diagnosis. IEEE Trans Instrum Meas 71:3503913. https://doi.org/10.1109/tim.2021.3139660
21. Tian YH (2020) Artificial intelligence image recognition method based on convolutional neural network algorithm. IEEE Access 8:125731–125744. https://doi.org/10.1109/access.2020.3006097
22. Wang H, Li YF, Men TL, Li LS (2024) Physically interpretable wavelet-guided networks with dynamic frequency decomposition for machine intelligence fault prediction. IEEE Trans Syst Man Cybern-Syst 54(8):4863–4875. https://doi.org/10.1109/tsmc.2024.3389068
23. Wang LJ, Zhang CD, Zhu J, Xu FX (2022) Fault diagnosis of motor vibration signals by fusion of spatiotemporal features. Machines 10(4):246. https://doi.org/10.3390/machines10040246
24. Xie SC, Tan HC, Yang CX, Yan HY (2023) A review of fault diagnosis methods for key systems of the high-speed train. Appl Sci-Basel 13(8):4790. https://doi.org/10.3390/app13084790
25. Yan RQ, Gao RX, Chen XF (2014) Wavelets for fault diagnosis of rotary machines: a review with applications. Signal Process 96:1–15. https://doi.org/10.1016/j.sigpro.2013.04.015

26. Ye Z, Yu JB (2021) Deep morphological convolutional network for feature learning of vibration signals and its applications to gearbox fault diagnosis. Mech Syst Sig Process 161:107984. https://doi.org/10.1016/j.ymssp.2021.107984
27. Yu D, Fu HY, Song YC, Xie WJ, Xie ZJ (2024) Deep transfer learning rolling bearing fault diagnosis method based on convolutional neural network feature fusion. Meas Sci Technol 35(1):015013. https://doi.org/10.1088/1361-6501/acfe31
28. Zhang Q, Deng LF (2023) An intelligent fault diagnosis method of rolling bearings based on short-time Fourier transform and convolutional neural network. J Fail Anal Prev 23(2):795–811. https://doi.org/10.1007/s11668-023-01616-9
29. Zhao HM, Liu J, Chen HY, Chen J, Li Y, Xu JJ, Deng W (2023) Intelligent diagnosis using continuous wavelet transform and Gauss convolutional deep belief network. IEEE Trans Reliab 72(2):692–702. https://doi.org/10.1109/tr.2022.3180273
30. Zhao L, Chi X, Li P, Ding JW (2023) Incipient fault feature enhancement of rolling bearings based on CEEMDAN and MCKD. Appl Sci-Basel 13(9):5688. https://doi.org/10.3390/app13095688
31. Zhong C, Wang JS, Sun WZ (2022) Fault diagnosis method of rotating bearing based on improved ensemble empirical mode decomposition and deep belief network. Meas Sci Technol 33(8):085109. https://doi.org/10.1088/1361-6501/ac6cc9
32. Zhou GZ, Guo ZQ, Sun SM, Jin QS (2023) A CNN-BiGRU-AM neural network for AI applications in shale oil production prediction. Appl Energy 344:121249. https://doi.org/10.1016/j.apenergy.2023.121249

Chapter 3
Ensemble Intelligent Diagnosis for Bearing Faults

3.1 Introduction

Intelligent transport systems are essential to modern industry and infrastructure of national economies [12, 18]. Over time, performance degradation in these systems can lead to equipment failures, potentially causing major safety incidents and substantial economic losses. Bearings, as key components of intelligent transport systems, are prone to failures, accounting for 30–40% of total failures [24]. The condition of the bearings significantly affects the operational efficiency and capacity of these systems, with severe faults potentially threatening system safety. Therefore, conducting research on bearing fault diagnosis is crucial for the timely and accurate detection of potential safety hazards in operation [17, 28]. This research is essential for ensuring the safe functioning of intelligent transport equipment, preventing major accidents, scheduling maintenance appropriately, and reducing unplanned downtime. The theoretical and practical of this research hold significant value in engineering applications.

The traditional fault diagnosis approach primarily utilizes signal processing techniques to extract fault characteristics and identifies the type of bearing fault based on empirical examples, as discussed in detail in Chapter 2. However, these methods, which rely on signal processing, present challenges [22, 29]. On one hand, they are too complex for analyzing massive datasets, and on the other hand, they lack the precision necessary to accurately determine the severity of faults.

Traditional intelligent diagnostics, leveraging artificial intelligence technologies such as SVM [19], ANN [9], ELM [21], and LSTM [10], are engineered to efficiently process and analyze large volumes of vibration data and automatically deliver diagnostic results. However, the learning capability of a single model is limited and cannot fully capture all fault information, which compromises the fault detection rate.

Ensemble learning is a potent method within deep learning, wherein mainstream approaches often integrate several weaker models to enhance overall performance [11, 30]. The effectiveness of this integration primarily hinges on the diversity of

H. Liu et al., *Prognostics and Health Management for Intelligent Electromechanical Systems*, https://doi.org/10.1007/978-981-96-7218-9_3

the models and the strategies used for integration. For instance, Li [14] introduced an innovative bearing fault diagnosis algorithm that employs an ensemble of deep CNN. Feng [8] developed an ensemble learning method tailored for classifying failure modes and predicting bearing capacity. Chen [5] designed an explainable deep ensemble model for bearing fault diagnosis, which not only capitalizes on feature learning capabilities but also enhances model interpretability. Additionally, Li [15] suggested a reinforcement ensemble method to diagnose rolling bearing faults.

This chapter focuses on preprocessing of vibration signals from the CWRU dataset to conduct an ensemble intelligent diagnosis on ten different bearing states. It utilizes two common ensemble methods: a multi-objective grey wolf optimizer-based model and a Boosting-based model. Finally, several evaluation metrics are established to assess the performance of different algorithms within the ensemble intelligent diagnosis framework.

3.2 Data Acquisition

In this chapter, experiments on bearing fault diagnosis are carried out with the high-frequency CWRU dataset [2], which is elaborated in Sect. 2.2. The experiments are carried out with 1772 r/min, 1 hp and 12 kHz. Time-domain waveform diagrams for ten types of bearing signals are displayed in Fig. 3.1. The same data augmentation techniques used in the previous chapter are employed here, involving the segmentation of samples through repeated superposition, with total length $L \approx 120{,}000$, sample length $l = 2048$, and a displacement D set to 1191, resulting in 100 samples per fault type.

3.3 Ensemble Diagnostic Model Based on Multi-objective Grey Wolf Optimizer for Bearing Faults

3.3.1 The Theoretical Basis of Empirical Wavelet Transform

The fundamental concept of empirical wavelet transform involves adaptively partitioning the spectrum of a Fourier-transformed signal to construct suitable orthogonal wavelet filters [23]. These filters are used to extract various amplitude-modulated and frequency-modulated components. Subsequent Hilbert transformation [27] of these components yields valuable instantaneous frequencies and amplitudes. Additionally, the corresponding Hilbert spectra are obtained. The process flow is depicted in Fig. 3.2.

Step 1: Based on the Shannon criterion, the fault signal is subjected to a Fast Fourier Transform (FFT), with the Fourier spectrum limited to the range of [0, 1].

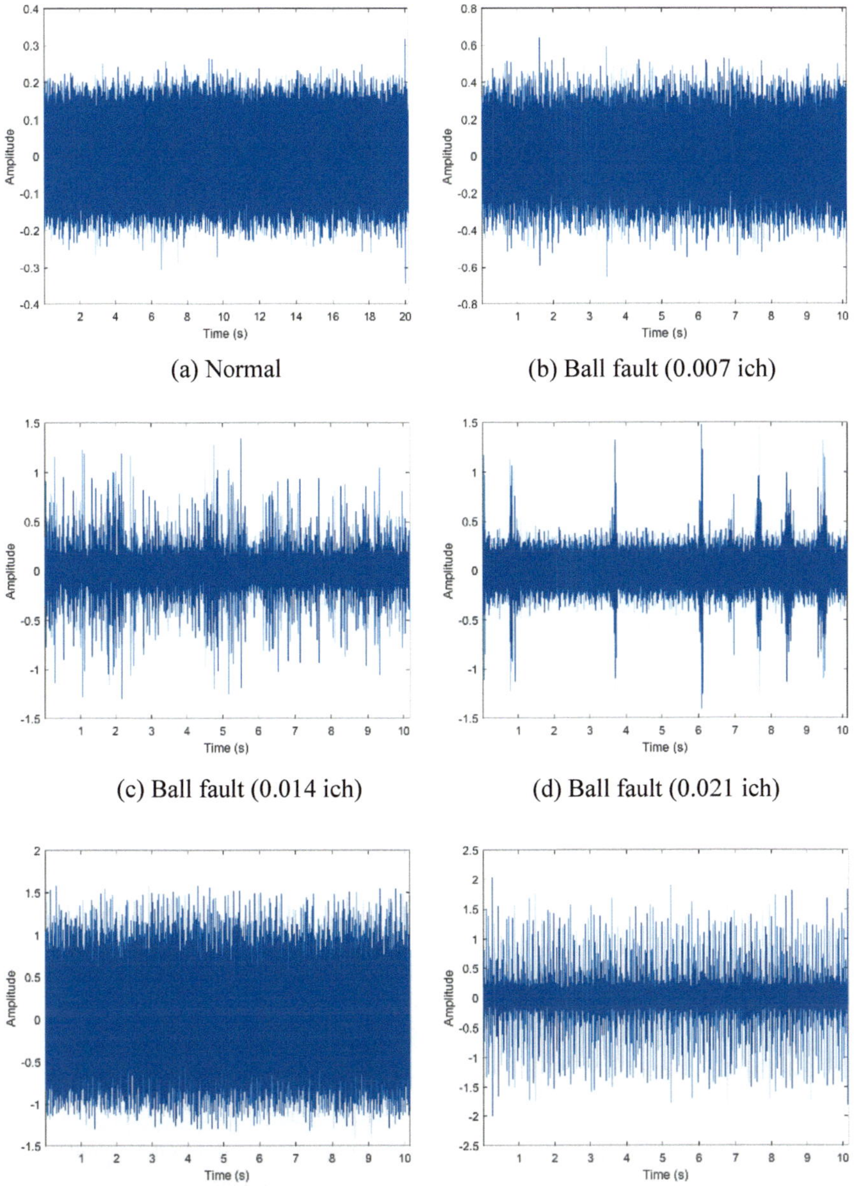

(a) Normal
(b) Ball fault (0.007 ich)
(c) Ball fault (0.014 ich)
(d) Ball fault (0.021 ich)
(e) InnerRace fault (0.007 ich)
(f) InnerRace fault (0.014 ich)

Fig. 3.1 Time-domain waveforms for 10 rolling bearing types

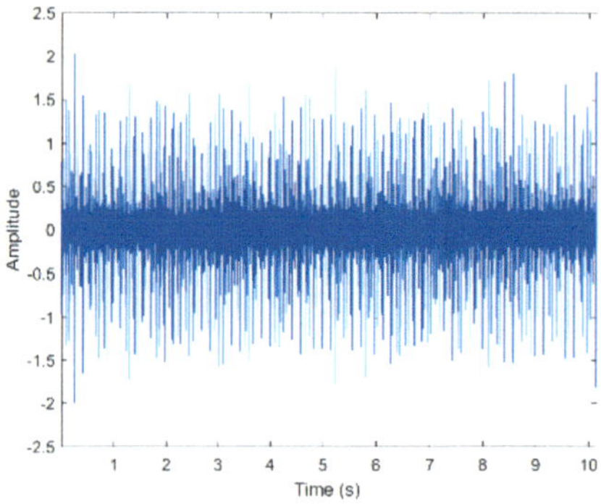

(g) InnerRace fault (0.021 ich)

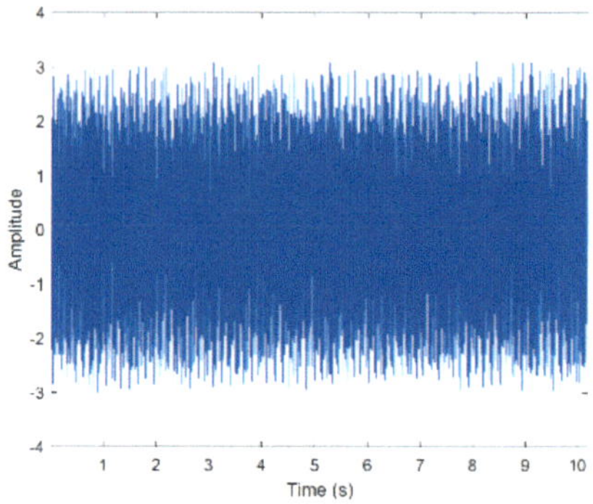

(h) OuterRace fault (0.007 ich)

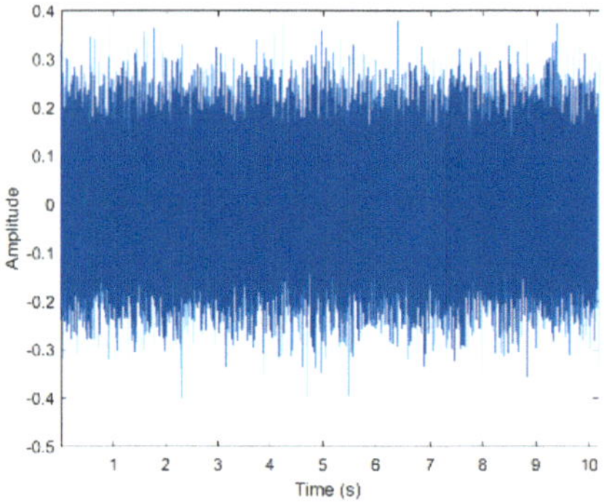

(i) OuterRace fault (0.014 ich)

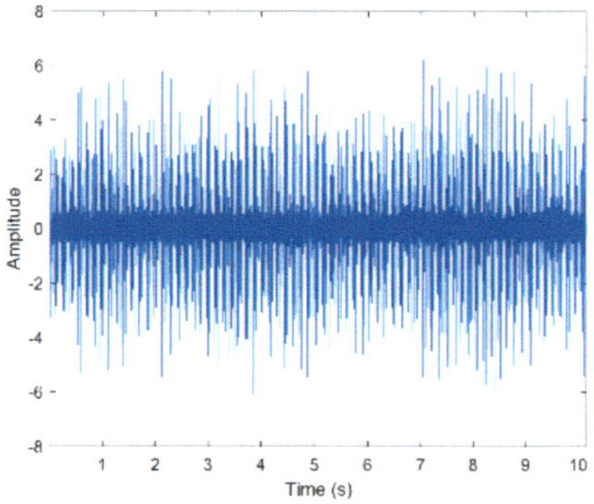

(j) OuterRace fault (0.021 ich)

Fig. 3.1 (continued)

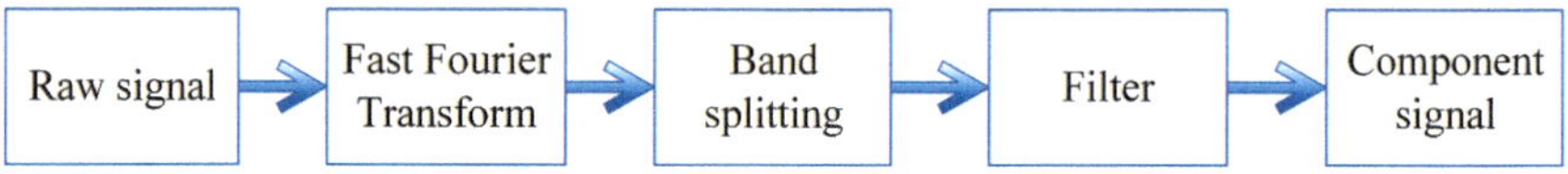

Fig. 3.2 The flowchart of EWT

Step 2: The Fourier spectrum is adaptively segmented to divide the signal into N single-component parts. If $M \geq N$, only the top N peaks are retained; if $M < N$, all peaks are kept and N is adjusted accordingly.
Step 3: The Fourier spectrum is divided into N continuous segments. If the midpoint between two consecutive peaks, w_n, is used as the boundary for each segment, then each frequency band can be represented as $\Lambda_n - [w_{n-1}, w_n]$, where $n = 1, 2, \ldots, N$, and $w_0 = 0$, $w_n = \pi$.
Step 4: According to Meyer theory, a wavelet window is applied to the segmented intervals.

Step 5: Constructing the detail coefficients and approximation coefficients of the EWT.

3.3.2 *The Theoretical Basis of Random Tree*

Random Forest (RF), as the name suggests, involves constructing a "forest" composed of multiple decision trees using randomized methods. It is an ensemble algorithm based on decision trees, which combined the Bagging ensemble learning theory with the random subspace method to develop this machine learning approach [13, 20]. The randomness in RF operates on two levels: first, by randomly selecting samples from the dataset, which diversifies each predictor's training subset and reduces the correlation between base predictors; second, by randomly selecting subsets of features at each node, where a fixed-size subset is chosen, and the most important feature within this subset is used for splitting. This dual randomness enhances the model's generalization performance and robustness, making RF well-suited for a variety of data types and task scenarios. The steps are outlined as follows (Fig. 3.3):

Step 1: Select a training set containing N samples, with each sample consisting of M features.
Step 2: Randomly select k features ($k \ll M$) as the candidate set of features for splitting at each decision tree node.
Step 3: For each node, select the optimal feature from the candidate feature set for splitting, such that the resulting impurity (e.g., information entropy) is minimized. The calculation formula is:

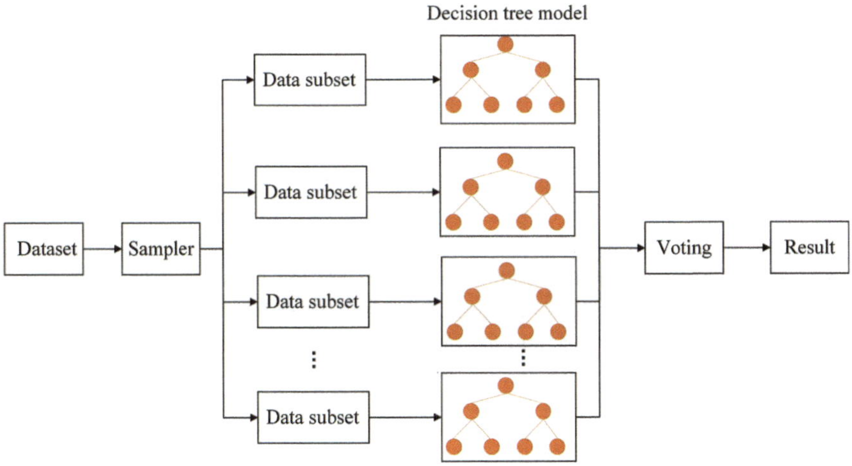

Fig. 3.3 Random forest structure

$$H(X) = -\sum_{i=1}^{n} p_i \log_2 p_i \quad (3.1)$$

where p_i denotes the probability distribution for each value of the random variable X.

Step 4: Repeat steps 2 and 3 to construct T decision trees.

Step 5: For a new test sample, input it into each decision tree to obtain a set of predictions. Then, aggregate these predictions using methods such as averaging or majority voting to produce the final prediction.

3.3.3 *The Theoretical Basis of Multi-objective Grey Wolf Optimizer*

The Multi-Objective Grey Wolf Optimizer (MOGWO) [16] enhances the original GWO [7] to support multi-objective optimization. It introduces two significant modifications compared to GWO:

The first change is the introduction of an archive [26]. Within this archive, the storage controller plays a crucial role, performing two key functions: first, it searches for the latest solutions; second, it manages the archiving process when the storage unit reaches its capacity limit. During the retrieval process, due to storage limitations in the archive, the algorithm needs to compare each newly generated solution with the existing ones after every iteration and then update the archive accordingly.

The second change is the introduction of a new leader selection strategy. The α, β, and δ wolves represent the top three best solutions and are responsible for guiding the other search agents in the global search for optimal solutions. However, in multi-objective optimization problems, it is challenging to directly compare solutions based on Pareto dominance to determine global optimality. Additionally, since the current optimal solutions are stored in the archive, the MOGWO algorithm uses a roulette wheel selection method. The chance of choosing each individual decreases as the number of individuals in the archive increases. The specific formula is as follows:

$$P_i = \frac{C}{N_i} \quad (3.2)$$

where C is a constant greater than 1. N_i represents the number of Pareto optimal solutions obtained in the ith storage unit. Additionally, unlike GWO, MOGWO does not distinguish between α, β, and δ wolves. Instead, these are collectively referred to as the "head wolves."

3.3.4 Experimental Result and Analysis

RF, EWT-RF, MOGWO-RF, and MOGWO-EWT-RF models are established to predict bearing fault diagnosis. The MOGWO algorithm optimizes the number and depth of the trees in the RF model, with the optimization range set to [1, 1000]. The results for the four models are presented in Table 3.1 and Figs. 3.4, 3.5, 3.6 and 3.7. Clearly, the MOGWO-EWT-RF model demonstrates superior performance across all evaluation metrics. The detailed analysis is as follows:

(1) The RF model shows the weakest. For instance, in terms of Acc, the prediction performance of RF is lower by 23.06%, 28.05%, and 39.12% compared to EWT-RF, MOGWO-RF, and MOGWO-EWT-RF, respectively. This is due to the complex patterns in the dataset that single-layer decision trees in the RF model struggle to capture effectively.
(2) EWT-RF shows a significant improvement over plain RF. For example, in terms of macro_P, EWT-RF increases by 32.14% compared to RF. This improvement is attributed to the wavelet transform, which better captures intrinsic patterns, thereby improving the efficacy of the RF model.
(3) MOGWO-RF also demonstrates a significant performance boost. For instance, in terms of macro_R, MOGWO-RF shows a 40.00% increase compared to RF. This improvement is due to the optimization provided by MOGWO, which helps better tune the RF model to focus on crucial features while also improving the balance between precision and recall.
(4) The best-performing model is MOGWO-EWT-RF, which improves macro_F1 by 62.50%, 24.66%, and 16.67% compared to RF, EWT-RF, and MOGWO-RF, respectively. This superior performance is due to the combination of hyperparameter tuning and enhanced feature extraction, resulting in better outcomes across all metrics. This approach proves especially effective for bearing fault diagnosis.

Table 3.1 Bearing fault diagnosis prediction and classification results

Model	Acc	macro_P	macro_R	macro_F1
RF	0.554	0.560	0.550	0.560
EWT-RF	0.720	0.740	0.720	0.730
MOGWO-RF	0.770	0.790	0.770	0.780
MOGWO-EWT-RF	0.910	0.910	0.910	0.910

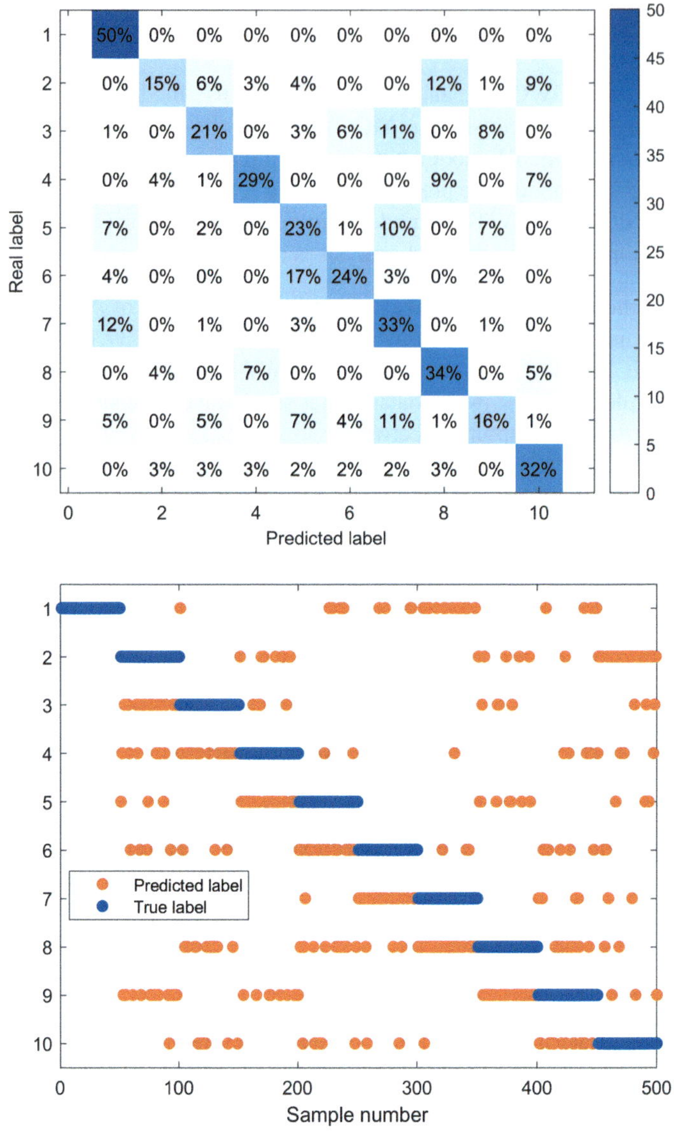

Fig. 3.4 RF fault classification results

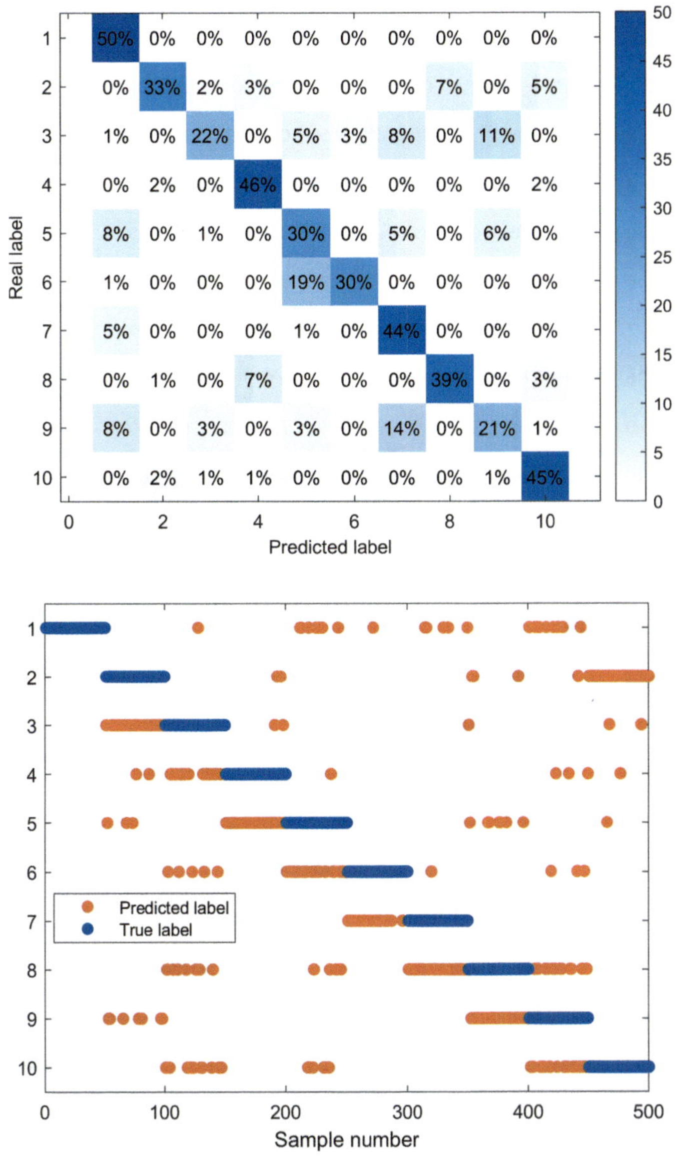

Fig. 3.5 EWT-RF fault classification results

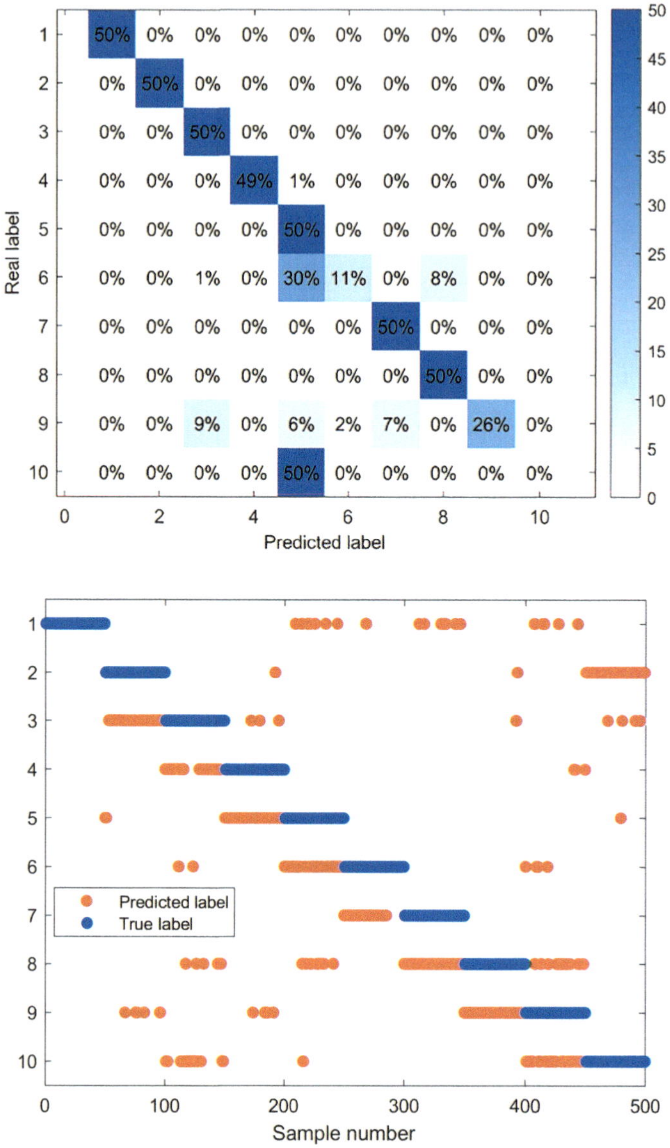

Fig. 3.6 MOGWO-RF fault classification results

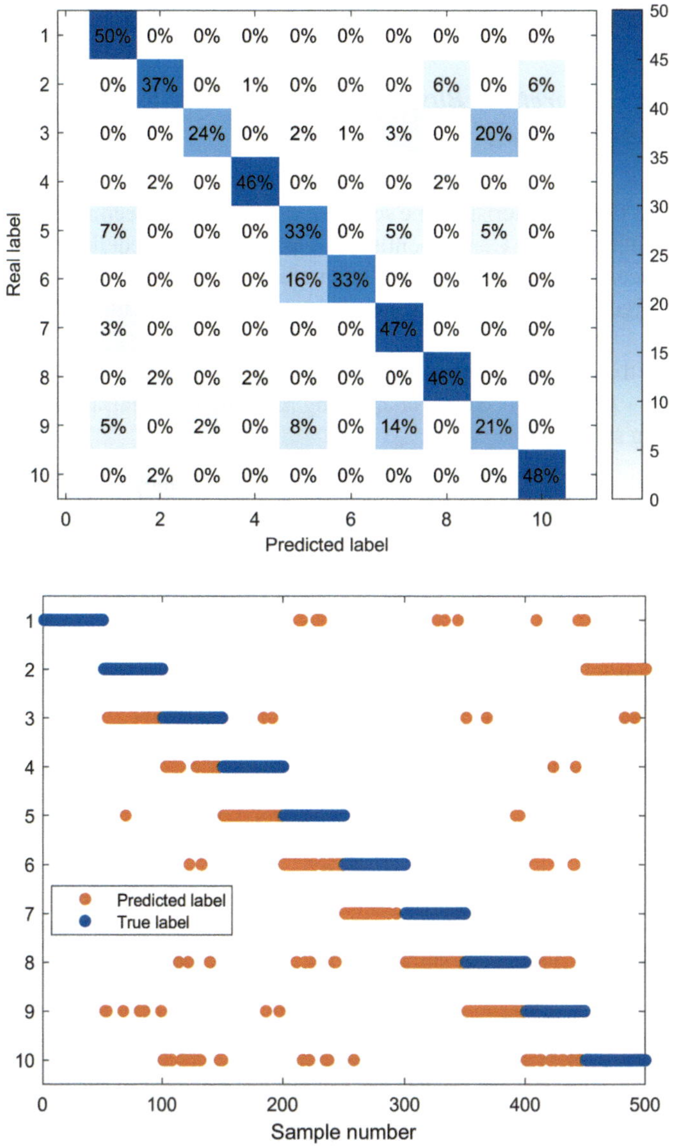

Fig. 3.7 MOGWO-EWT-SVM fault classification results

3.4 Boosting Ensemble Diagnostic Model for Bearing Faults

3.4.1 The Theoretical Basis of Empirical Mode Decomposition

EMD utilizes the characteristic time scales of data for signal decomposition, thereby offering adaptability [3]. Consequently, there is no need to predefine basis functions in signal processing. Therefore, EMD decomposition theoretically can be applied to any type of signal, offering clear advantages in handling non-stationary and nonlinear data.

Consider the one-dimensional signal $f(n)$ as a discrete function defined on $F = \{0, 1, \ldots, N-1\}$. The procedure for applying EMD to the initial signal $f(n)$ includes the following steps:

Step 1: Identify all the local maxima and minima of the $f(n)$. Use cubic spline interpolation to fit these points, creating the upper envelope $e_{max}(n)$ and the lower envelope $e_{min}(n)$. It is important to ensure that all extremum points are encompassed.
Step 2: Calculate the mean $m_1(n)$ of $e_{max}(n)$ and $e_{min}(n)$.
Step 3: To obtain the first component $h_1(n)$, subtract the mean envelope $m_1(n)$ from the initial signal $f(n)$. If $h_1(n)$ serves as the first-order IMF of the $f(n)$, move to **Step 4**. Otherwise, repeat **Steps 1** and **2**.
Step 4: Subtracting the first order IMF function $C_1(n)$ from the initial signal $f(n)$ to obtain the residual $r_1(n)$.

$$r_1(n) = f(n) - C_1(n) \tag{3.3}$$

Step 5: Use the residual $r_1(n)$ as the new signal and repeat Step 1 through 4 to derive a subsequent residual $r_2(n)$. Continue this process until the termination criteria are met. The initial signal $f(n)$ decomposed by EMD can be calculated as:

$$f(n) = \sum_{i=1}^{I} C_i(n) + r_I(n) \tag{3.4}$$

Following the steps above, the EMD decomposition results for signals from 10 types of rolling bearings in a motor running at 1772 r/min are shown in Fig. 3.8.

3.4.2 The Theoretical Basis of Boosting

Boosting ensembles, also known as boosting methods, involve a strong dependency among base learners, thus constituting a serial algorithm model [1]. The algorithm

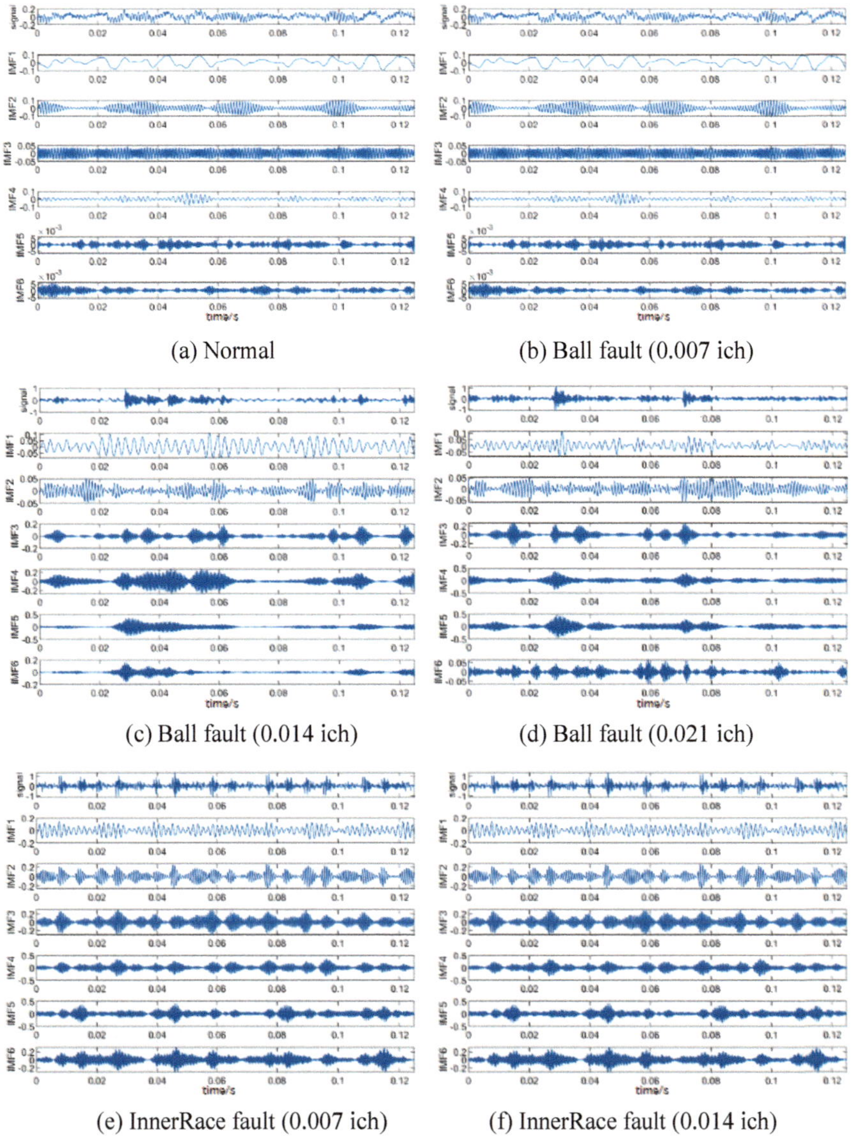

Fig. 3.8 Results of EMD decomposition of bearing vibration signals

aims to reduce the bias of base learners by utilizing the prediction results of the previous base learner to assign higher weights to incorrectly predicted samples for subsequent learning. By continuously increasing the weights of hard-to-predict samples, it enhances the base learners' ability to recognize them. This iterative training continues until the stopping criteria are met. Subsequently, the predictions of

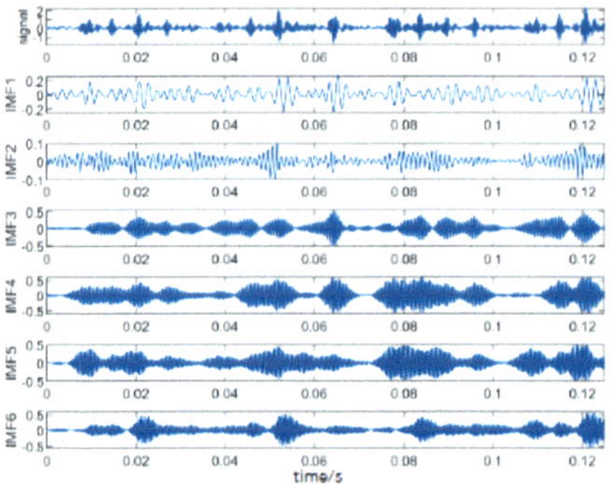

(g) InnerRace fault (0.021 ich)

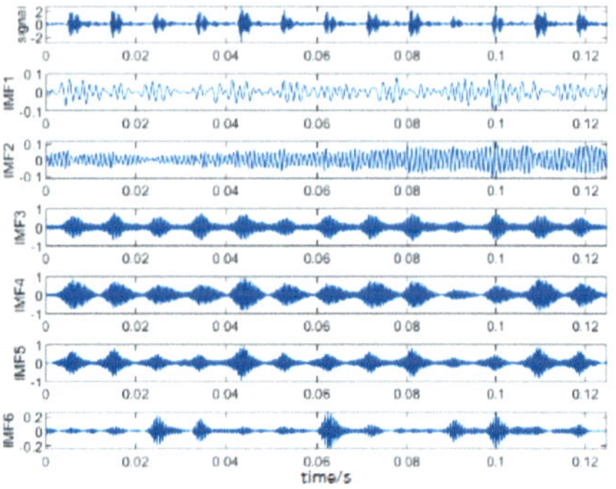

(h) OuterRace fault (0.007 ich)

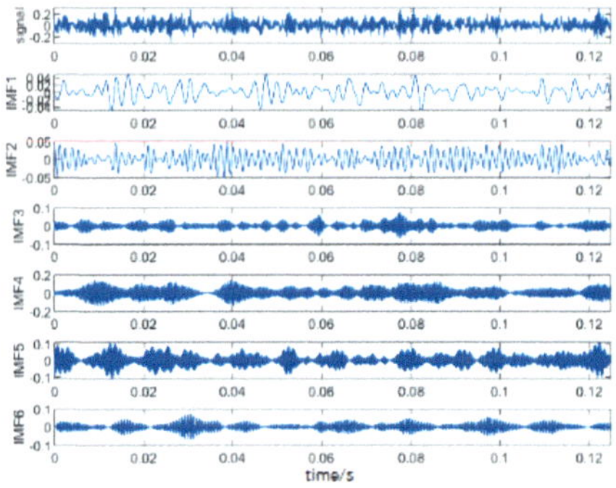

(i) OuterRace fault (0.014 ich)

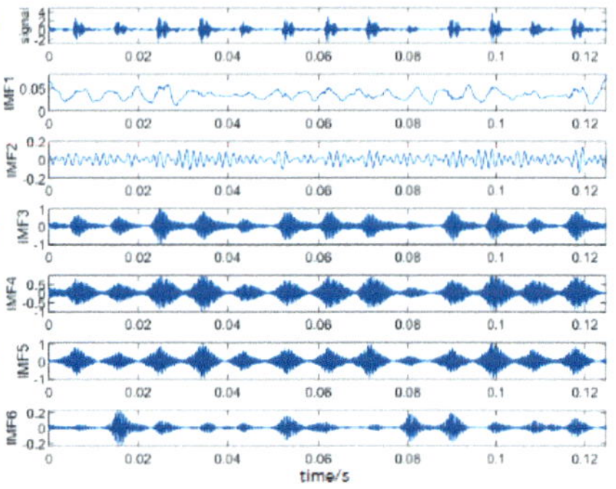

(j) OuterRace fault (0.021 ich)

Fig. 3.8 (continued)

various base learners are combined through voting or a weight-based linear combination to make the final prediction. The framework of the Boosting ensemble is displayed in Fig. 3.9.

Initially, the sample weights in the training set are set, typically to $1/m$. First, base learner 1 is trained, and samples misclassified by base learner 1 are assigned higher weights, updating the training set. Subsequently, base learner 2 is trained, followed by the serial training of N base learners in sequence. The test set is then input into each of the N-base learners, and all prediction results are combined through voting or a weight-based linear combination to produce the final prediction. The Boosting algorithm combines N weak learners into a strong learner, reducing the classification error of the strong learner gradually as long as each weak learner's accuracy exceeds 50%. Boosting algorithms include LogitBoost, AdaBoost, among others. In this section, AdaBoost is utilized.

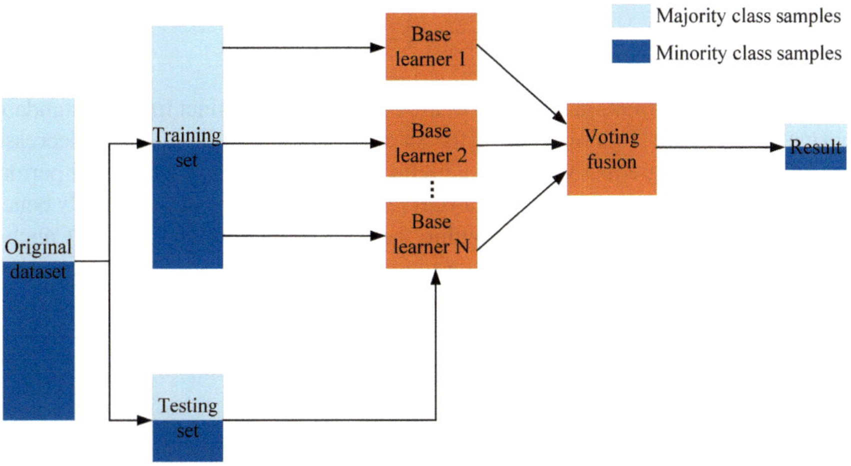

Fig. 3.9 Framework diagram of boosting

3.4.3 The Theoretical Basis of the Osprey-Cauchy-Sparrow Search Algorithm

The Osprey-Cauchy-Sparrow search algorithm (OCSSA) combines the hunting behavior of ospreys, the long-tail characteristics of the Cauchy distribution, and the social competition behavior of sparrows [4]. The mathematical model of SSA can be found in [25]. The repetition process, flowchart, and pseudocode of OOA are detailed in [6] and are not repeated in this book. The logistic chaotic map, a typical representation of complex nonlinear behavior, is also known as the parabolic map. Its formula is as follows:

$$x_{k+1} = \mu x_k(1 - x_k) \tag{3.5}$$

where x_k is a randomly selected number within the range (0, 1], and μ, which lies within the interval (0, 4], serves as a tunable parameter. The value of x_{k+1} is determined by mapping x_k. For the purposes of this study, μ is set to 3.98 to guarantee that the initial sequence is uniformly distributed.

In SSA, scroungers typically gather near the most efficient forager and might sometimes compete for food to take over as the producer. To avoid local optima and improve global optimization, Cauchy mutation strategy is integrated into the position update formula for scroungers. The updated formula for the scroungers' positions is outlined below:

$$X_{i,j}^{k+1} = X_{best}(k) + Cauchy(0, 1) \oplus X_{best}(k) \tag{3.6}$$

where $\oplus$ is used to indicate multiplication. $Cauchy(0, 1)$ is defined as follows:

$$f(x) = \frac{1}{\pi}\left(\frac{1}{x^2 + 1}\right), \ -\infty < x < \infty \tag{3.7}$$

The Cauchy distribution shares similarities but remains distinct from the standard normal distribution. It is recognized for its central peak and fat tails that decrease towards zero more gradually, making it highly effective for introducing larger perturbations than those typical of a Gaussian curve. This quality proves particularly beneficial for optimization algorithms. Within the SSA framework, employing Cauchy mutations to update the positions of sparrows substantially expands the search capabilities of the algorithm. Crucially, this mutation strategy increases the SSA's ability to circumvent local optima by promoting broader exploration and aiding in overcoming suboptimal solutions. The steps of OCSSA are depicted in the flowchart shown in Fig. 3.10.

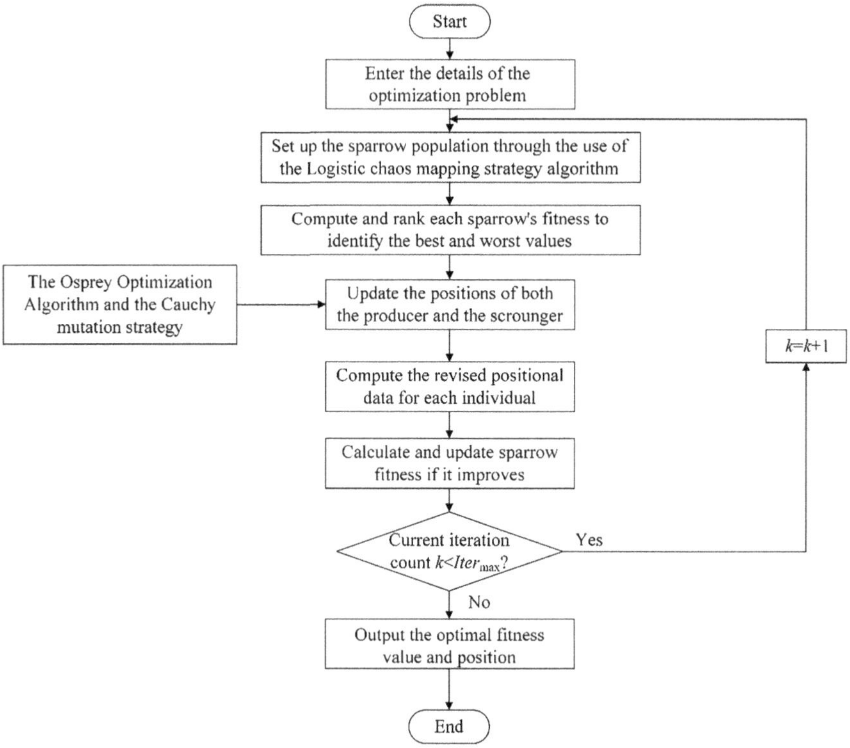

Fig. 3.10 Flowchart of the OCSSA

3.4.4 Experimental Result and Analysis

Boosting, EMD-Boosting, and OCSSA-EMD-Boosting are developed to predict bearing fault diagnosis. The optimization settings for OCSSA encompass the maximum sifting iterations and the decomposition modes for EMD, with their respective optimization ranges specified as [50, 1000] for iterations and [3, 10] for modes. The optimization goal is the sample entropy. The outcomes of the three models are presented in Table 3.2 and Figs. 3.11, 3.12 and 3.13. It is clear that MOGWO-EWT-RF has achieved the best results across all performance metrics, with a detailed analysis as follows:

(1) Boosting exhibits the weakest performance across all metrics. For instance, compared to EMD-Boosting and OCSSA-EMD-Boosting, Boosting's classification prediction performance is lower by 34.60% and 38.32%, respectively. This subpar performance may be attributed to issues such as overfitting when handling complex bearing vibration signals, inadequate data preprocessing, or improper feature selection.
(2) Incorporating EMD into Boosting notably enhances its effectiveness. Taking macro_P and macro_R as examples, the performance has improved by 60.34% and 32.54% respectively compared to standard Boosting. This improvement is due to the preprocessing of data via EMD, which equips the boosting algorithm to better manage data variations and complex patterns. This enhancement in noise reduction and feature extraction contributes significantly to increases in accuracy and other related metrics.
(3) OCSSA further improves the EMD-Boosting model, resulting in the highest performance among the three. For instance, macro_F1 is increased by 66.97% and 6.74% compared to Boosting and EMD-Boosting, respectively. This improvement is due to OCSSA's ability to strike an effective equilibrium between global and local search capabilities. By optimizing the EMD parameters, OCSSA helps in more accurately extracting fault-related IMFs, thereby enhancing the precision of fault diagnosis.

Table 3.2 Prediction and classification results of bearing fault diagnosis

	Acc	macro_P	macro_R	macro_F1
Boosting	0.586	0.580	0.590	0.570
EMD-Boosting	0.896	0.930	0.900	0.890
OCSSA-EMD-Boosting	0.950	0.950	0.950	0.950

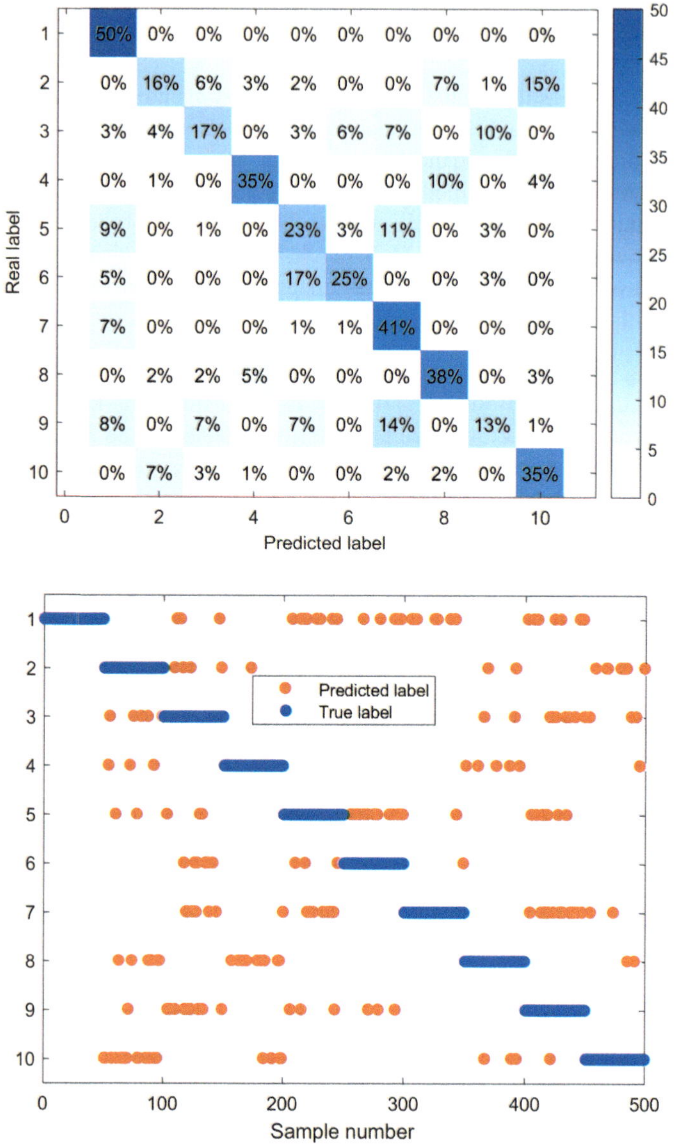

Fig. 3.11 Boosting fault classification results

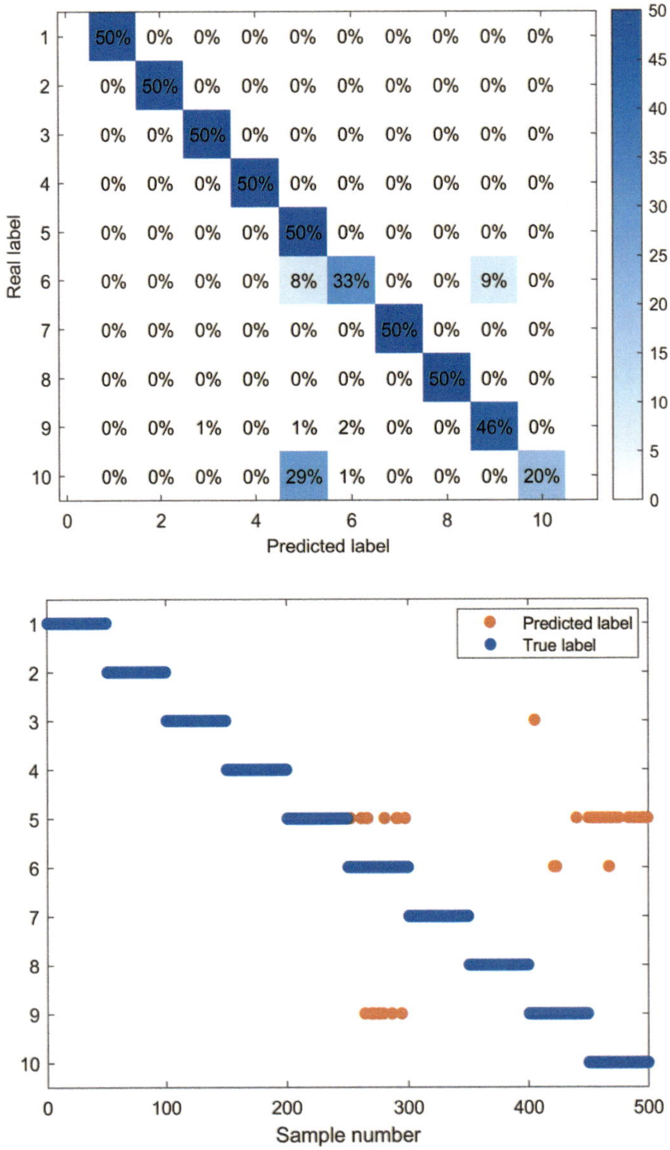

Fig. 3.12 EMD-Boosting fault classification results

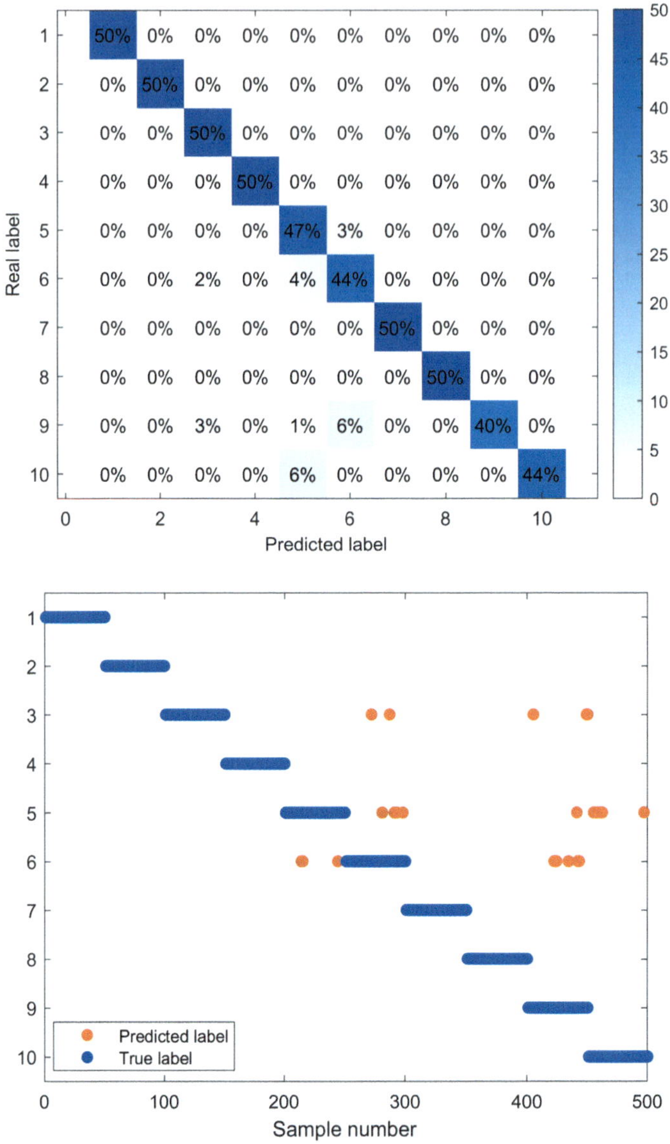

Fig. 3.13 OCSSA-EMD-Boosting fault classification results

3.5 Model Performance Comparison

In this chapter, the applicability of the optimized MOGWO-based model and the Boosting-based model in bearing fault diagnosis is analyzed. Ten sets of bearing vibration signal data are chosen, and comprehensive fault diagnosis prediction experiments are conducted for each group. The optimization in this chapter focuses on two aspects: feature extraction and hyperparameter optimization. Based on the predictive experiments, the comparison of various evaluation metrics is illustrated in Fig. 3.14.

The model performance comparison can be summarized as follows:

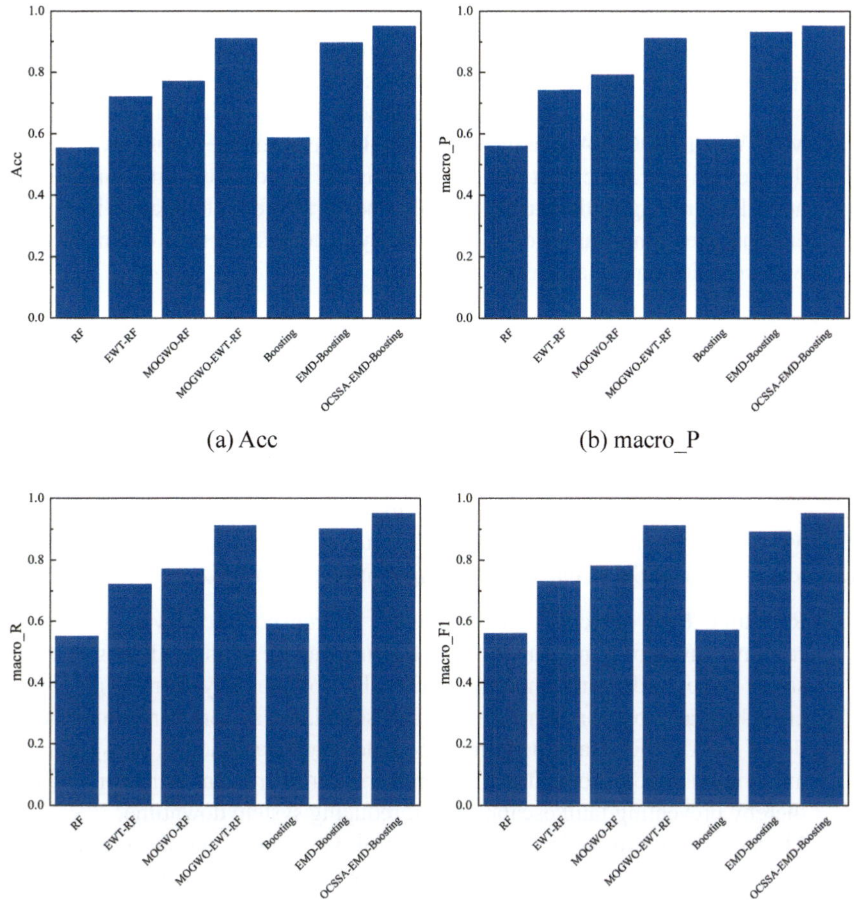

(a) Acc (b) macro_P

(a) macro_R (b) F1

Fig. 3.14 Model performance comparison of all metrics

(1) Individual ensemble intelligent models, including RF and Boosting, have limited accuracy due to issues such as noise interference, non-stationarity, and insufficient feature extraction. However, Boosting, with its strategy of progressively improving weak classifiers, may be more effective in handling complex patterns or noisy bearing vibration signal data. This approach places higher demands on the predictor, making it a necessary consideration.
(2) After embedding the EWT and EMD algorithms into the RF-based and Boosting-based models, respectively, the accuracy of both models improved to varying degrees. This indicates that effective feature capture, along with the ability to process and analyze key features extracted from complex, nonlinear, and non-stationary bearing vibration signals, can markedly improve the predictive accuracy.
(3) By using MOGWO to optimize the RF model's hyperparameters and OCSSA for the VMD in the Boosting-based model, the accuracy of both models has improved to varying degrees. Utilizing these advanced optimization algorithms, MOGWO and OCSSA, to adjust the critical hyperparameters not only enhances the predictive accuracy but also strengthens the models' reliability and robustness. This highlights the crucial role of hyperparameter optimization in modern ensemble learning frameworks, particularly in addressing fault diagnosis tasks with complex data characteristics.

3.6 Conclusions

Bearing fault diagnosis is critical for PHM in intelligent transportation systems such as high-speed trains, autonomous vehicles, and aircraft. Bearings, as common key components in these systems, directly impact the overall reliability and safety. Below are some potential applications of integrated intelligent bearing fault diagnosis technology in the PHM of intelligent transportation systems:

(1) Real-time fault detection and response: In transportation systems such as railways, aviation, and automotive, bearing failures can lead to significant safety issues and operational disruptions. By continuously monitoring parameters such as vibration, temperature, and sound, the system can automatically detect and diagnose fault patterns. Upon identifying anomalies, the system can autonomously adjust operational parameters or trigger maintenance procedures, thereby preventing fault escalation and reducing system downtime.
(2) Maintenance optimization and cost control: Continuous monitoring and fault diagnosis of bearings allow for more accurate predictions of maintenance needs, enabling condition-based maintenance rather than traditional scheduled maintenance. This approach can significantly reduce unnecessary maintenance tasks and associated costs, while also extending bearing life and improving overall system efficiency.
(3) Integrated intelligent decision support: Bearing fault diagnosis systems can be integrated with other intelligent systems, such as dynamic scheduling systems, smart grid management, and emergency response systems, to collectively

enhance overall operational intelligence. For instance, in a smart grid, if a bearing in the cooling system of an electric vehicle charging station malfunctions, the system can automatically adjust charging power or reallocate power resources in real time to prevent overload and improve efficiency.

(4) Lifecycle management: Integrated intelligent bearing fault diagnosis technology can provide real-time data and predictive analytics to ensure optimal performance and reliability throughout the equipment's lifecycle. For example, in high-speed rail systems, a lifecycle-based bearing management system not only monitors health during the usage phase but also collects and analyzes data at every stage, from design and manufacturing to decommissioning. This approach optimizes operational efficiency, enhances sustainability and cost-effectiveness, and ensures safe, reliable, and efficient service for passengers.

References

1. Almahdi A, Al Mamlook RE, Bandara N, Almuflih AS, Nasayreh A, Gharaibeh H, Alasim F, Aljohani A, Jamal A (2023) Boosting ensemble learning for freeway crash classification under varying traffic conditions: a hyperparameter optimization approach. Sustainability 15(22):15896. https://doi.org/10.3390/su152215896
2. Althobiani F (2024) A novel framework for robust bearing fault diagnosis: preprocessing, model selection, and performance evaluation. IEEE Access 12:59018–59036. https://doi.org/10.1109/access.2024.3390234
3. Centeno-Bautista MA, Rangel-Rodriguez AH, Perez-Sanchez AV, Amezquita-Sanchez JP, Granados-Lieberman D, Valtierra-Rodriguez M (2023) Electrocardiogram analysis by means of empirical mode decomposition-based methods and convolutional neural networks for sudden cardiac death detection. Appl Sci-Basel 13(6):3569. https://doi.org/10.3390/app13063569
4. Chang Y, Bao GQ (2024) Enhancing rolling bearing fault diagnosis in motors using the OCSSA-VMD-CNN-BiLSTM model: a novel approach for fast and accurate identification. IEEE Access 12:78463–78479. https://doi.org/10.1109/access.2024.3408628
5. Chen ZY, Qin W, He GL, Li JP, Huang RY, Jin G, Li WH (2023) Explainable deep ensemble model for bearing fault diagnosis under variable conditions. IEEE Sens J 23(15):17737–17750. https://doi.org/10.1109/jsen.2023.3281505
6. Dehghani M, Trojovský P (2023) Osprey optimization algorithm: a new bio-inspired meta-heuristic algorithm for solving engineering optimization problems. Front Mech Eng 8:1126450. https://doi.org/10.3389/fmech.2022.1126450
7. Faris H, Aljarah I, Al-Betar MA, Mirjalili S (2018) Grey wolf optimizer: a review of recent variants and applications. Neural Comput Appl 30(2):413–435. https://doi.org/10.1007/s00521-017-3272-5
8. Feng DC, Liu ZT, Wang XD, Jiang ZM, Liang SX (2020) Failure mode classification and bearing capacity prediction for reinforced concrete columns based on ensemble machine learning algorithm. Adv Eng Inf 45:101126. https://doi.org/10.1016/j.aei.2020.101126
9. Gunerkar RS, Jalan AK, Belgamwar SU (2019) Fault diagnosis of rolling element bearing based on artificial neural network. J Mech Sci Technol 33(2):505–511. https://doi.org/10.1007/s12206-019-0103-x
10. Hao SJ, Ge FX, Li YM, Jiang JY (2020) Multisensor bearing fault diagnosis based on one-dimensional convolutional long short-term memory networks. Measurement 159:107802. https://doi.org/10.1016/j.measurement.2020.107802
11. Hasegawa T, Kondo K (2023) Easy ensemble: simple deep ensemble learning for sensor-based human activity recognition. IEEE Internet Things J 10(6):5506–5518. https://doi.org/10.1109/jiot.2022.3222221

12. Jiao YT, Wang J, Li RM, Wang FY, Zhao HT, Xiong G (2024) Intelligent generation of test cases for a parallel testing system: a case study on railway systems. IEEE Intell Transp Syst Mag 1–12. https://doi.org/10.1109/mits.2024.3435828
13. Karabadji NE, Korba AA, Assi A, Seridi H, Aridhi S, Dhifli W (2023) Accuracy and diversity-aware multi-objective approach for random forest construction. Expert Syst Appl 225:120138. https://doi.org/10.1016/j.eswa.2023.120138
14. Li SB, Liu GK, Tang XH, Lu JG, Hu JJ (2017) An ensemble deep convolutional neural network model with improved D-S evidence fusion for bearing fault diagnosis. Sensors 17(8):1729. https://doi.org/10.3390/s17081729
15. Li YN, Zhang Y, Wang RX, Fu JF (2024) A reinforcement ensemble learning method for rolling bearing fault diagnosis under variable work conditions. Sensors 24(11):3323. https://doi.org/10.3390/s24113323
16. Mirjalili S, Saremi S, Mirjalili SM, Coelho LD (2016) Multi-objective grey wolf optimizer: a novel algorithm for multi-criterion optimization. Expert Syst Appl 47:106–119. https://doi.org/10.1016/j.eswa.2015.10.039
17. Peng C, Sheng YY, Gui WH, Tang ZH, Li CY (2024) A rolling bearing fault diagnosis method based on multimodal knowledge graph. IEEE Trans Ind Inf 20(11):13047–13057. https://doi.org/10.1109/tii.2024.3431074
18. Song TR, Schonfeld P, Pu H (2023) A review of alignment optimization research for roads, railways and rail transit lines. IEEE Trans Intell Transp Syst 24(5):4738–4757. https://doi.org/10.1109/tits.2023.3235685
19. Song XM, Wei WH, Zhou JB, Ji GJ, Hussain G, Xiao MH, Geng GS (2023) Bayesian-optimized hybrid kernel SVM for rolling bearing fault diagnosis. Sensors 23(11):5137. https://doi.org/10.3390/s23115137
20. Sun ZG, Wang GT, Li PF, Wang H, Zhang M, Liang XW (2024) An improved random forest based on the classification accuracy and correlation measurement of decision trees. Expert Syst Appl 237:121549. https://doi.org/10.1016/j.eswa.2023.121549
21. Udmale SS, Singh SK (2019) Application of spectral kurtosis and improved extreme learning machine for bearing fault classification. IEEE Trans Instrum Meas 68(11):4222–4233. https://doi.org/10.1109/tim.2018.2890329
22. Wang C, Gan M, Zhu CA (2017) Intelligent fault diagnosis of rolling element bearings using sparse wavelet energy based on overcomplete DWT and basis pursuit. J Intell Manuf 28(6):1377–1391. https://doi.org/10.1007/s10845-015-1056-2
23. Wang D, Zhao Y, Yi C, Tsui KL, Lin JH (2018) Sparsity guided empirical wavelet transform for fault diagnosis of rolling element bearings. Mech Syst Sig Process 101:292–308. https://doi.org/10.1016/j.ymssp.2017.08.038
24. Wang M, Yu J, Leng HY, Du XS, Liu YR (2024) Bearing fault detection by using graph autoencoder and ensemble learning. Sci Rep 14(1):5206. https://doi.org/10.1038/s41598-024-55620-6
25. Xue JK, Shen B (2020) A novel swarm intelligence optimization approach: sparrow search algorithm. Syst Sci Control Eng 8(1):22–34. https://doi.org/10.1080/21642583.2019.1708830
26. Yang YF, Yang B, Wang SL, Jin TG, Li S (2020) An enhanced multi-objective grey wolf optimizer for service composition in cloud manufacturing. Appl Soft Comput 87:106003. https://doi.org/10.1016/j.asoc.2019.106003
27. Zalik B, Strnad D, Podgorelec D, Kolingerová I, Lukac L, Lukac N, Kolmanic S, Zalik KR, Kohek S (2023) A new transformation technique for reducing information entropy: a case study on greyscale raster images. Entropy 25(12):1591. https://doi.org/10.3390/e25121591
28. Zhang JS, Zhang K, An YY, Luo H, Yin S (2024) An integrated multitasking intelligent bearing fault diagnosis scheme based on representation learning under imbalanced sample condition. IEEE Trans Neural Networks Learn Syst 35(5):6231–6242. https://doi.org/10.1109/tnnls.2022.3232147
29. Zhang M, Yin J, Chen WL (2022) Rolling bearing fault diagnosis based on time-frequency feature extraction and IBA-SVM. IEEE Access 10:85641–85654. https://doi.org/10.1109/access.2022.3198701
30. Zhang QS, Tsang ECC, He Q, Guo YT (2023) Ensemble of kernel extreme learning machine based elimination optimization for multi-label classification. Knowl-Based Syst 278:110817. https://doi.org/10.1016/j.knosys.2023.110817

Chapter 4
Deep Learning Prediction for Bearing Remaining Useful Life

4.1 Introduction

Currently, bearings play a crucial role in intelligent operation and maintenance systems [21]. As these systems evolve towards greater digitalization and intelligence, the level of automation in maintenance equipment is continuously increasing [9]. For example, in high-speed trains, bearings are essential support components in the running gear of locomotives and rolling stock. he condition, dynamic performance, and longevity of bearings are vital to maintain the safe, stable, and efficient operation of trains [11]. Bearing failures in the running gear during operation can lead to severe issues such as axle seizure, axle burning, or axle breakage, which may result in equipment failure and even pose safety risks to both passengers and vehicles [7, 11]. Additionally, the maintenance and repair efforts required are substantial, leading to significant losses. Therefore, research into bearing life prediction methods holds substantial practical value.

Methods to predicting the Remaining Useful Life (RUL) of bearings can generally be divided into model-based methods, machine learning models, and deep learning techniques [13, 24]. For instance, Gazizulin [4] developed a physical failure model using ABAQUS to predict the RUL of equipment, taking into account the current operating state and equipment load. Nonetheless, developing a physical model that accurately captures the degradation trend necessitates a thorough comprehension of the system's failure mechanisms. As equipment becomes increasingly complex, it is often challenging to fully grasp these mechanisms, or the cost of obtaining this information may be prohibitively high.

Cutting-edge intelligent information technologies are injecting new momentum into the field of intelligent operation and maintenance. On one hand, machine learning algorithms, including Logistic Regression [23], Linear Regression [3], Support Vector Regression [12, 20], and RVM [8], have achieved significant success. On the other hand, many researchers have introduced deep learning methods into the study of bearing RUL prediction. Typical approaches include autoencoder networks

H. Liu et al., *Prognostics and Health Management for Intelligent Electromechanical Systems*, https://doi.org/10.1007/978-981-96-7218-9_4

[25], deep belief networks [16], CNN [15], and RNN [6]. Among these, RNNs have shown strong performance in RUL prediction. However, RNNs are prone to gradient explosion and vanishing gradient issues. In response, Wang [22] derived feature-weighted scores and inputted the highest-scoring features into an LSTM model for predicting bearing RUL. Cheng [2] combined CNN and BiLSTM for assessing the health of rolling bearings. Que [19] constructed a stacked GRU to predict bearing RUL. Numerous studies have demonstrated that deep learning algorithms possess significant advantages and performance in health monitoring and intelligent operation and maintenance of industrial components. Therefore, researching bearing RUL prediction algorithms based on deep learning methods has become a key area of focus.

In this chapter, the PHM 2012 dataset is utilized to forecast the RUL of bearings across three distinct operating conditions. Two commonly used deep learning methods are applied: BiLSTM-based model and GRU-based model. Finally, evaluation metrics such as Mean Absolute Error (MAE), Mean Absolute Percentage Error (MAPE), and Root Mean Square Error (RMSE) are established to assess the performance of different algorithms.

4.2 Data Acquisition

Since RUL prediction requires full lifecycle data of bearings as well as data collected under actual operating conditions. Given that gathering such comprehensive data typically requires a long period, this chapter utilizes the PHM 2012 dataset [14]. This dataset was obtained by experimental methods that accelerated bearing degradation, such as increasing the load and rotational speed, reducing the degradation time to just a few hours. This approach addresses the challenges of data collection, where obtaining such data is difficult and time-consuming (Table 4.1).

Table 4.1 IEEE PHM 2012 dataset description

Case	Load (N)	Rotate speed (r/min)	Sample size
Case 1	4000	1800	2803
			871
			1802
			1139
			2302
			2302
			1502

(continued)

Table 4.1 (continued)

Case	Load (N)	Rotate speed (r/min)	Sample size
Case 2	4200	1650	911
			797
			1955
			751
			2311
			701
			120
Case 3	5000	1500	515
			1637
			434

The accelerated full lifecycle experiments on the rolling bearings of the PRONOSTIA test rig were conducted under three different operating conditions, each corresponding to a distinct load and rotational speed. Two types of sensors—speed sensors and temperature sensors—were utilized for measurement. The speed sensors were mounted on the bearing, with data collected every 10 s in each direction and recorded for 0.1 s at 25.6 kHz. Since the horizontal data exhibits more pronounced lifecycle characteristic changes, the experimental data presented in this chapter is derived from measurements taken in the horizontal direction.

The first two sets of data from conditions 1, 2, and 3 are selected as the training set, and Bearing1_3, Bearing2_3, and Bearing3_3 are chosen for the testing set. The time-domain waveform diagrams for the training and testing sets selected in the vertical direction are displayed in Figs. 4.1 and 4.2, respectively.

The method for calculating the life ratio T is shown in Eq. (4.1):

$$T = \frac{RUL_t}{RUL_f} \tag{4.1}$$

where RUL_f represents the actual full lifespan of the bearing. F In the PHM2012 dataset, Bearing 1_1 accumulated 2803 datasets from the beginning of operation to failure. Considering the data collection interval for PHM2012 is every 10 s, the total operational lifespan of Bearing 1_1 amounts to 28,030 s. RUL_t represents the RUL of the bearing at time t. This yields a life ratio in the range of (0, 1).

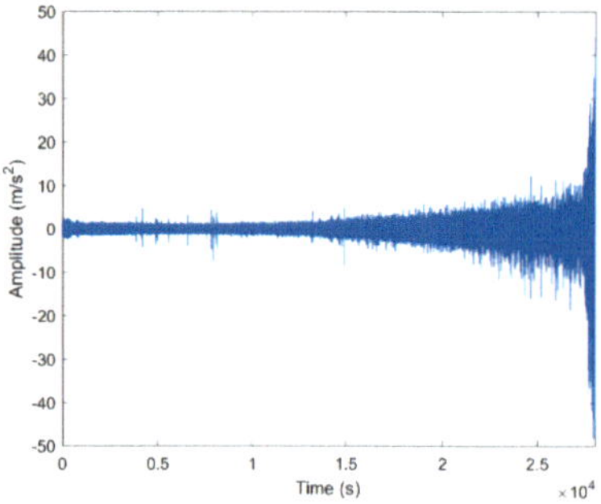

(a) Vertical vibration signal (Bearing 1_1)

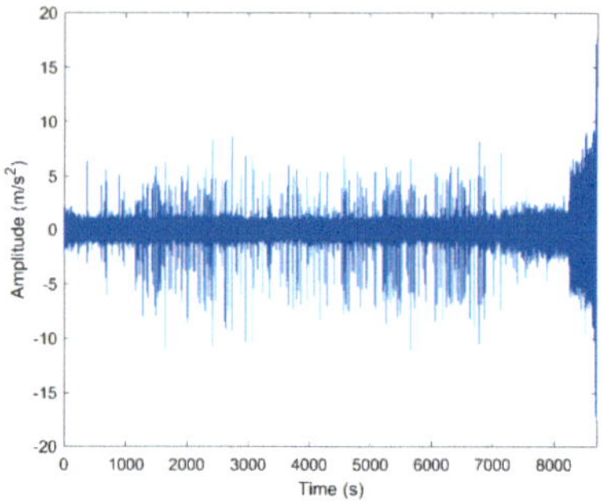

(b) Vertical vibration signal (Bearing 1_2)

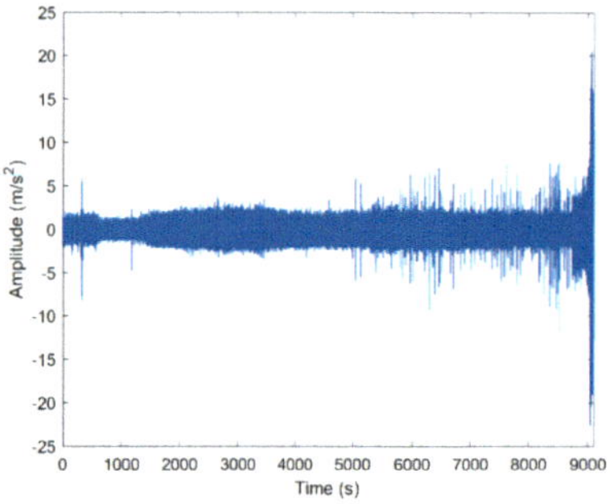

(c) Vertical vibration signal (Bearing 2_1)

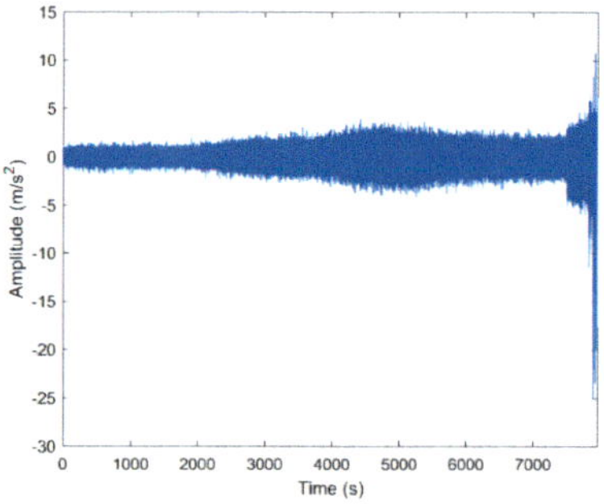

(d) Vertical vibration signal (Bearing 2_2)

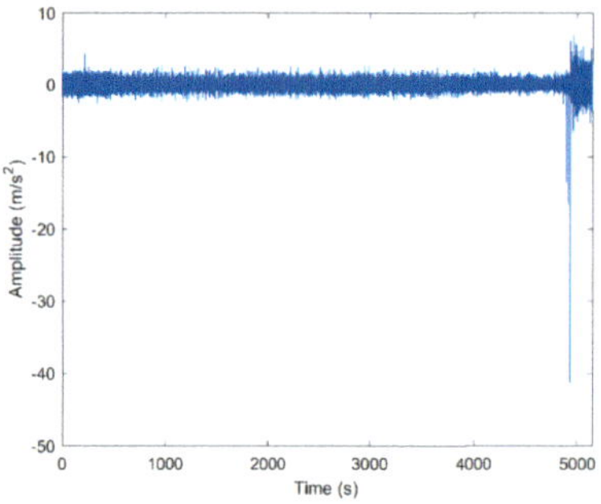

(e) Vertical vibration signal (Bearing 3_1)

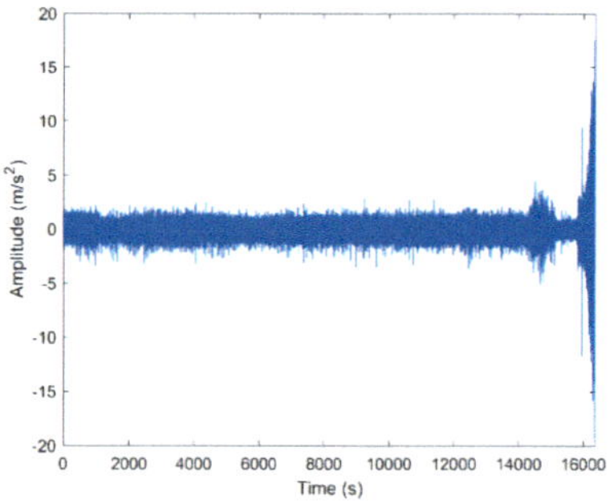

(f) Vertical vibration signal (Bearing 3_2)

Fig. 4.1 Time-domain waveform diagram of the training set

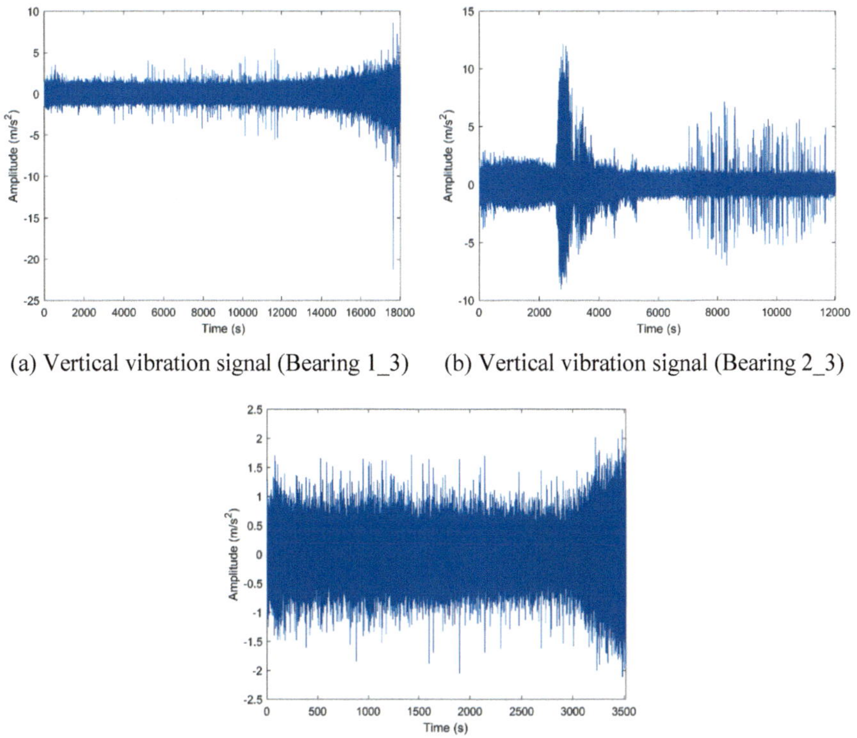

(a) Vertical vibration signal (Bearing 1_3) (b) Vertical vibration signal (Bearing 2_3)

(c) Vertical vibration signal (Bearing 3_3)

Fig. 4.2 Time-domain waveform diagram of the testing set

4.3 BiLSTM-Based Predictive Model for Bearing Remaining Useful Life

4.3.1 The Theoretical Basis Convolutional Neural Network

The CNN employs convolutional operations and is fundamental to deep learning technologies [27]. These networks include a feature extraction module comprised of convolutional and pooling layers. Within the convolutional layer, connections between neurons are limited to neighboring subsets, which minimizes the overall network complexity. Specifically, for bearing signal data, a one-dimensional convolution architecture is implemented, where convolutional kernel size is set to 3.

Once the convolution computation is finalized, the outputs from each layer undergo non-linear transformation via an activation function. Popular activation functions include Sigmoid, Tanh, and Relu. For the purposes of this analysis, the Relu activation function is selected to adjust the outputs of the convolutional layers.

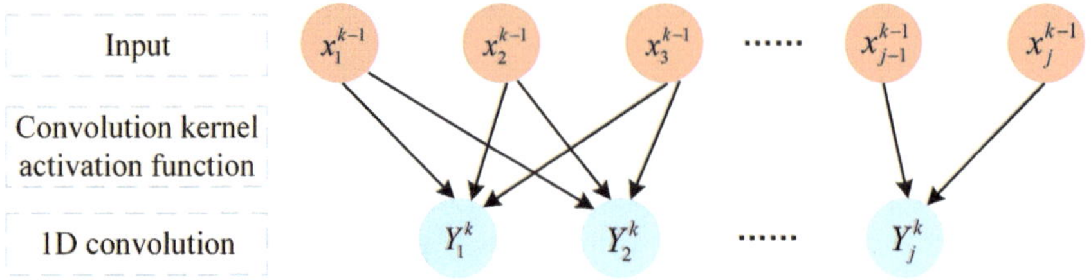

Fig. 4.3 Illustration of the one-dimensional convolution operation

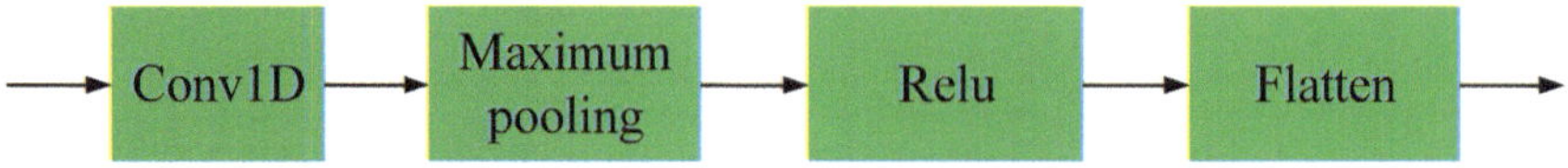

Fig. 4.4 The architecture of CNN

$$Y_j^k = \mathrm{Relu}\left(b_j^k + \sum_{i=1}^{N_{k-1}} conv1D\left(x_i^{k-1}, w_{ij}^{k-1}\right)\right) \tag{4.2}$$

where x_i^{k-1} is the input from layer k-1, w_{ij}^{k-1} denotes the convolution kernel, and b_j^k is the bias or threshold value of layer k; b_j^k constitutes the set of input features. The operation *conv1D* is applied to execute one-dimensional convolution, as shown in Fig. 4.3.

The pooling layer in the network reduces the parameter count. Incorporating a pooling layer enhances computational speed and helps avoid overfitting [28]. In this study, max pooling is utilized to effectively lower the data's dimensionality. The architecture of the CNN is depicted in Fig. 4.4.

$$X = \max(x_1, x_2, x_3, \ldots, x_n) \tag{4.3}$$

where x_i represents each value pooled, and X is the result of the maximum pooling operation.

After passing through various convolutional layers, activation functions, and pooling layers, the features Y are extracted from the input data. These features are subsequently input into a BiLSTM for prediction

$$Y = \mathrm{Relu}\left(\mathrm{Relu}\left(\cdots \mathrm{Relu}\left({}_j^k + \sum_{i=1}^{N_{k-1}} conv1D\left(x_i^{k-1}, w_{ij}^{k-1}\right)\right)\right)\right) \tag{4.4}$$

4.3.2 *The Theoretical Basis Bidirectional Long Short-Term Memory*

BiLSTM merges a forward-propagating LSTM with a backward-propagating LSTM network [5, 10]. This bidirectional approach enables the effective utilization of the features of energy consumption time series data by processing past and future energy data independently, thereby enhancing the precision of predictions related to energy consumption. Figure 4.5 illustrates the architecture of the BiLSTM.

Unlike traditional LSTM, BiLSTM processes energy consumption data from both past and future directions simultaneously within the time series. The forward LSTM segment predicts future energy consumption based on past data, while the backward LSTM segment models past energy consumption using future data. The bidirectional processing capability of BiLSTM efficiently captures and extracts the relationships between energy consumption data and its influencing factors. The output from the BiLSTM is detailed below:

$$\begin{cases} \vec{h}_t = \overrightarrow{\mathrm{LSTM}}\left(\vec{h}_{t-1}, x_t\right), \ t \in [1, T] \\ \overleftarrow{h}_t = \overleftarrow{\mathrm{LSTM}}\left(\overleftarrow{h}_{t-1}, x_t\right), \ t \in [T, 1] \end{cases} \tag{4.5}$$

$$y_t^b = \mu \vec{h}_t + \upsilon \overleftarrow{h}_t \tag{4.6}$$

where and $\vec{h}_t$, $\overleftarrow{h}_t$ represent the RUL outputs from the forward and backward LSTM, respectively. $\vec{h}_{t-1}$ and $\overleftarrow{h}_{t-1}$ is the RUL outputs of the bearing from the forward and backward LSTM at time $t - 1$. The variable y_t^b is the predicted RUL output of the bearing by the BiLSTM. The forward factor μ and the backward factor υ are determined through training on the bearing vibration signals.

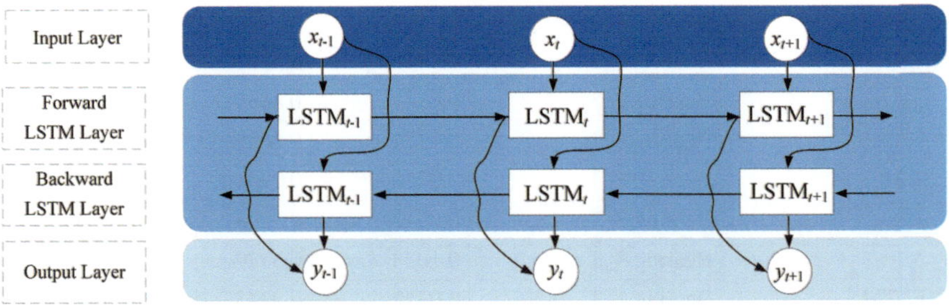

Fig. 4.5 Structure of the BiLSTM network

4.3.3 Experimental Result and Analysis

RNN, LSTM, BiLSTM, and CNN-BiLSTM models are established to predict the bearings RUL under three different conditions. The results of the four models are presented in Table 4.2 and Fig. 4.6. Clearly, the CNN-BiLSTM model achieves the best performance across all evaluation metrics, with a detailed analysis as follows:

(1) The RNN model generally exhibits higher errors compared to LSTM, BiLSTM, and CNN-BiLSTM models. Taking the MAE of Bearing1_3 as an example, the RNN's performance worsens by 91.6%, 101.8%, and 350.47% respectively when compared to LSTM, BiLSTM, and CNN-BiLSTM. This is attributed to its simpler structure, which may not effectively capture the complex time dependencies present in bearing vibration signals.
(2) LSTM and BiLSTM offer significant improvements over RNN, as they are variations of the RNN. Taking the MAPE of Bearing2_3 as an example, LSTM and BiLSTM respectively show accuracy enhancements of 14.00% and 33.31% compared to RNN. This improvement is due to the architecture of LSTM and BiLSTM, which more effectively extracts key features from time series data. It is clear that BiLSTM performs better than LSTM, as it can analyze data across the entire usage cycle of the bearing (including past loads and future trends), thus more accurately predicting the point of bearing failure.
(3) CNN-BiLSTM demonstrates superior performance across all bearings and metrics. Taking the RMSE of Bearing3_3 as an example, the accuracy improvements over RNN, LSTM, and BiLSTM are 78.01%, 72.73%, and 71.33% respectively. This indicates the effectiveness of CNN layers in extracting relevant features from time series data, which are then sequentially processed through BiLSTM for enhanced predictive accuracy.

Table 4.2 The predictive performance of CNN-BiLSTM in all cases

Model	Bearing number	MAE	MAPE	RMSE
RNN	Bearing 1_3	0.1437	0.4219	0.1752
	Bearing 2_3	0.1541	0.4335	0.1836
	Bearing 3_3	0.1274	0.4005	0.1642
LSTM	Bearing 1_3	0.0750	0.2934	0.0967
	Bearing 2_3	0.0947	0.3728	0.1056
	Bearing 3_3	0.1028	0.3779	0.1324
BiLSTM	Bearing 1_3	0.0712	0.2375	0.0788
	Bearing 2_3	0.0685	0.2891	0.0814
	Bearing 3_3	0.0913	0.2356	0.1259
CNN-BiLSTM	Bearing 1_3	0.0319	0.1169	0.0365
	Bearing 2_3	0.0470	0.2115	0.0525
	Bearing 3_3	0.0316	0.1088	0.0361

Fig. 4.6 The predictive results of bearings in all cases

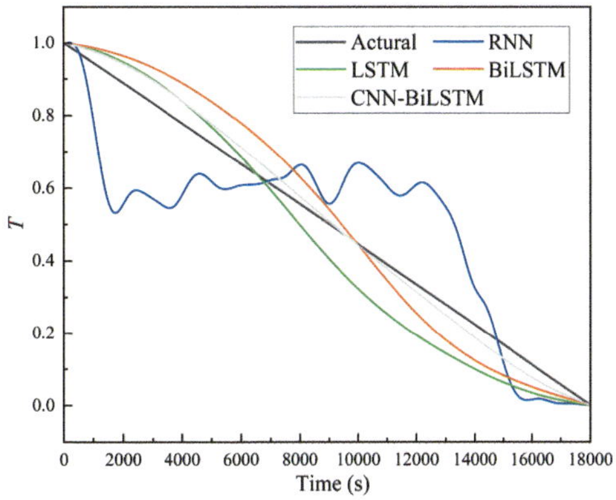

(a) Bearing1_3

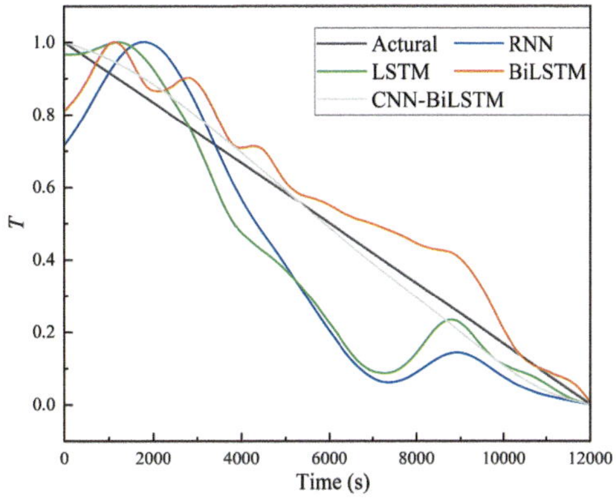

(b) Bearing2_3

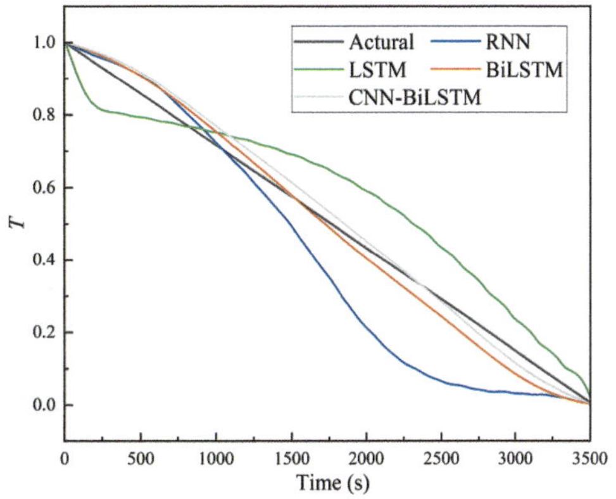

(c) Bearing3_3

4.4 GRU-Based Predictive Model for Bearing Remaining Useful Life

4.4.1 The Theoretical Basis Gate Recurrent Unit

The GRU, an adaptation of the LSTM design, effectively shortens training time while preserving the accuracy of predictions [1]. GRU simplifies the parameter count relative to LSTM by utilizing the update and reset gates [17]. The detailed structure of the GRU is illustrated in Fig. 4.7.

$$r_t = \sigma\left(W^{(r)}x_t + U^{(r)}h_{t-1}\right) \tag{4.7}$$

$$z_t = \sigma\left(W^{(z)}x_t + U^{(z)}h_{t-1}\right) \tag{4.8}$$

$$\tilde{h}_t = \tanh(r_t \times Uh_{t-1} + Wx_t) \tag{4.9}$$

$$h_t = (1 - z_t) \times h_{t-1} + z_t \times \tilde{h}_t \tag{4.10}$$

The forward propagation equation for the GRU is as follows:

Where x_t is the input to the network at time t. r_t and z_t represent the outputs of the reset gate and the update gate, respectively. h_t and h_{t-1} represent the hidden state of the next node and the hidden layer state at time $t - 1$, respectively. $\tilde{h}_t$ signifies the computation process involving the candidate set from input x_t and the output h_{t-1}. The σ and tanh are used as the sigmoid function and the activation function,

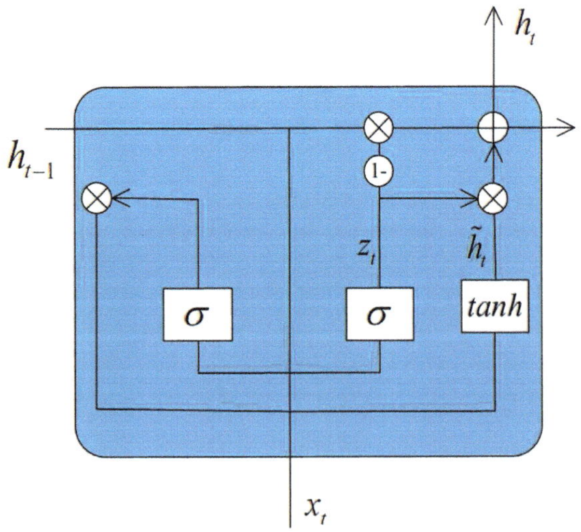

Fig. 4.7 Structure of the GRU network

respectively. The matrices $U^{(z)}$, $W^{(z)}$, $U^{(r)}$, $W^{(r)}$, U, and W are the weights involved in the network, with $\times$ indicating matrix multiplication.

4.4.2 The Theoretical Basis Attention

Attention, characterized as a targeted model for allocating probabilities, emphasizes the distribution of information and the influence of critical data to enhance prediction accuracy [18, 26]. The architecture of the GRU, refined through the attention mechanism, is depicted in Fig. 4.8.

Attention mechanism starts by pinpointing essential sequences within the input through a process of weighted summation, described as follows:

$$a_i = \frac{\exp(S_i)}{\sum_{j=1}^{T_x} \exp(S_j)} \tag{4.11}$$

$$S_i = \upsilon \tanh(Wh_i + b) \tag{4.12}$$

The Attention mechanism generates feature vectors by forming weighted aggregates of node vectors from the hidden layer, depicted as:

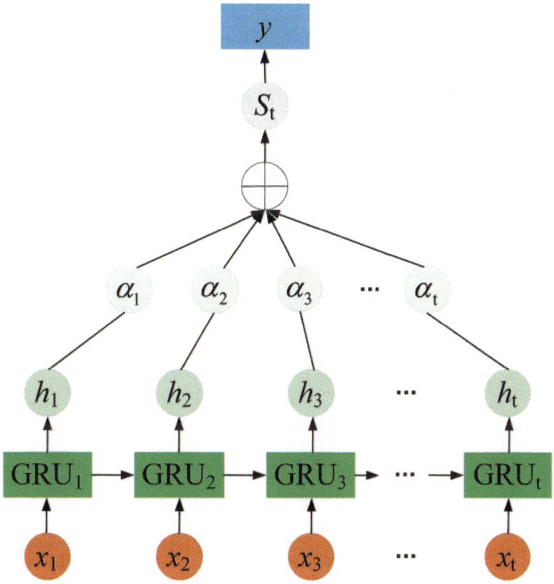

Fig. 4.8 The structure of GRU-Attention

$$C = \sum_{j=1}^{T_x} a_{ki} h_i \tag{4.13}$$

The final output reads:

$$y = H(C, h, x) \tag{4.14}$$

The Attention mechanism addresses the challenge of effectively capturing earlier historical information in the GRU network and amplifies the influence of critical information within the bearing vibration signal sequence.

4.4.3 Experimental Result and Analysis

Multilayer Perceptron (MLP), Attention, GRU, and GRU-Attention models are implemented to predict the bearings RUL under three different conditions. The outcomes of these models are depicted in Table 4.3 and Fig. 4.9. It is evident that the GRU-Attention model outperforms the others. A detailed analysis follows:

(1) GRU demonstrates significantly better performance across all metrics compared to Attention. For example, focusing on the RMSE of Bearing3_3, GRU decreased by 29.34% over Attention. This emphasizes GRU's strength in managing sequences and adeptly capturing time-dependent patterns in the bearing vibration signals.
(2) GRU-Attention exhibits the best performance among all tested models. Taking the RMSE of Bearing1_3 as an example, GRU-Attention shows a performance improvement of 59.70% over Attention and 29.10% over GRU. This superior performance is due to the GRU-Attention hybrid effectively leveraging the strengths of both GRU and the Attention mechanism. While the GRU maintains

Table 4.3 The predictive performance of GRU-Attention in all cases

Model	Bearing number	MAE	MAPE	RMSE
Attention	Bearing1_3	0.0665	0.2266	0.0744
	Bearing2_3	0.0677	0.2709	0.0784
	Bearing3_3	0.1304	0.3386	0.1687
GRU	Bearing1_3	0.0378	0.1302	0.0438
	Bearing2_3	0.0352	0.1577	0.0393
	Bearing3_3	0.0874	0.2331	0.1192
GRU-Attention	Bearing1_3	0.0268	0.1233	0.0306
	Bearing2_3	0.0290	0.1232	0.0328
	Bearing3_3	0.0304	0.1525	0.0361

Fig. 4.9 The predictive result of bearings in all cases

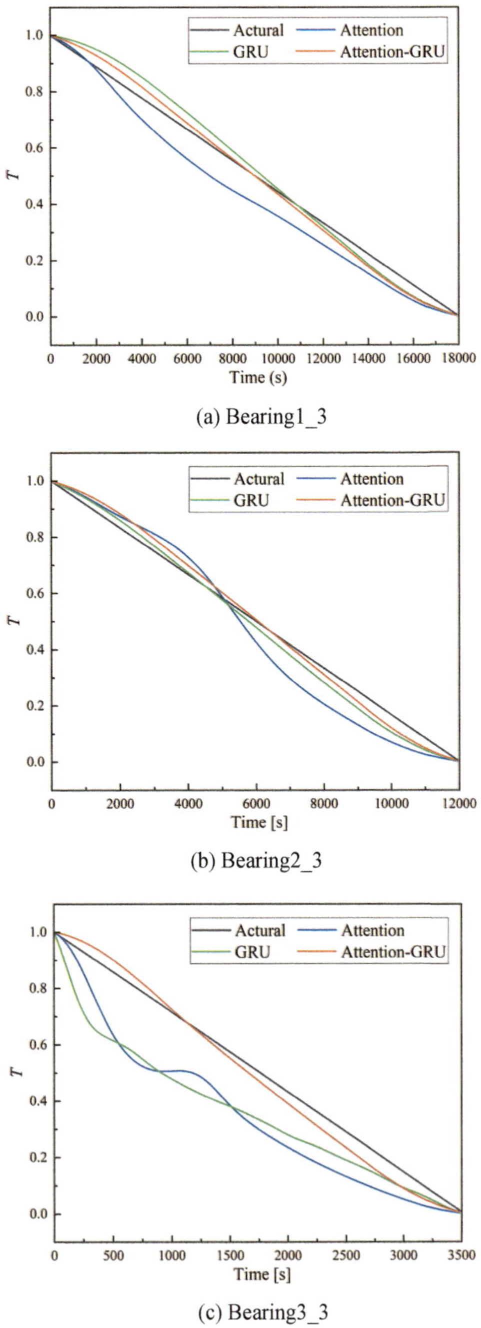

relevant information throughout the sequence, the Attention layer selectively concentrates on the most crucial segments of the sequence to enhance prediction accuracy.

4.5 Model Performance Comparison

In this chapter, two classic deep learning models are employed to assess their suitability for predicting the RUL of bearings. Three states of bearing vibration signal data are selected for analysis. Optimization in this chapter is approached from two angles: firstly, through deep learning feature extraction, and secondly, via Attention optimization. Based on the predictive experiments conducted, performance comparisons across various metrics are illustrated in Fig. 4.10, and the summary of model performance is as follows:

(1) The individual BiLSTM and GRU models outperform RNN, LSTM, and Attention in predicting the RUL of bearings. This enhanced performance can be attributed to various factors. Firstly, the BiLSTM model, by learning dependencies in both forward and backward directions, captures contextual information more comprehensively. This bidirectional learning capability enhances the BiLSTM's effectiveness in RUL prediction tasks. On the other hand, the GRU model, with its simplified structure, offers increased efficiency when processing longer sequences, which can significantly reduce training times and resource consumption.
(2) The integration of CNN and BiLSTM significantly enhances the performance of predicting the bearings RUL. This improvement stems from the complementary capabilities of CNN's powerful local feature extraction and BiLSTM's ability to capture temporal dependencies. CNN effectively extracts key features from raw data through its convolutional layers, while BiLSTM utilizes these features to accurately model both forward and backward dependencies in the time series. This approach not only optimizes the model's learning efficiency capability but also boosts prediction accuracy and robustness in complex and nonlinear data environments, making it particularly suitable for complex physical processes.
(3) Embedding Attention within GRU has further enhanced the accuracy of predicting bearings RUL. This enhancement is credited to the Attention mechanism, optimizing the flow of information and the feature decoding process. This combination leverages GRU's capability to capture long-term dependencies and Attention's adaptive focus, resulting in more precise predictions and increased model efficiency when dealing with complex time-series data. This approach is particularly suitable for the fine-grained health monitoring of electromechanical systems where detailed management is essential.

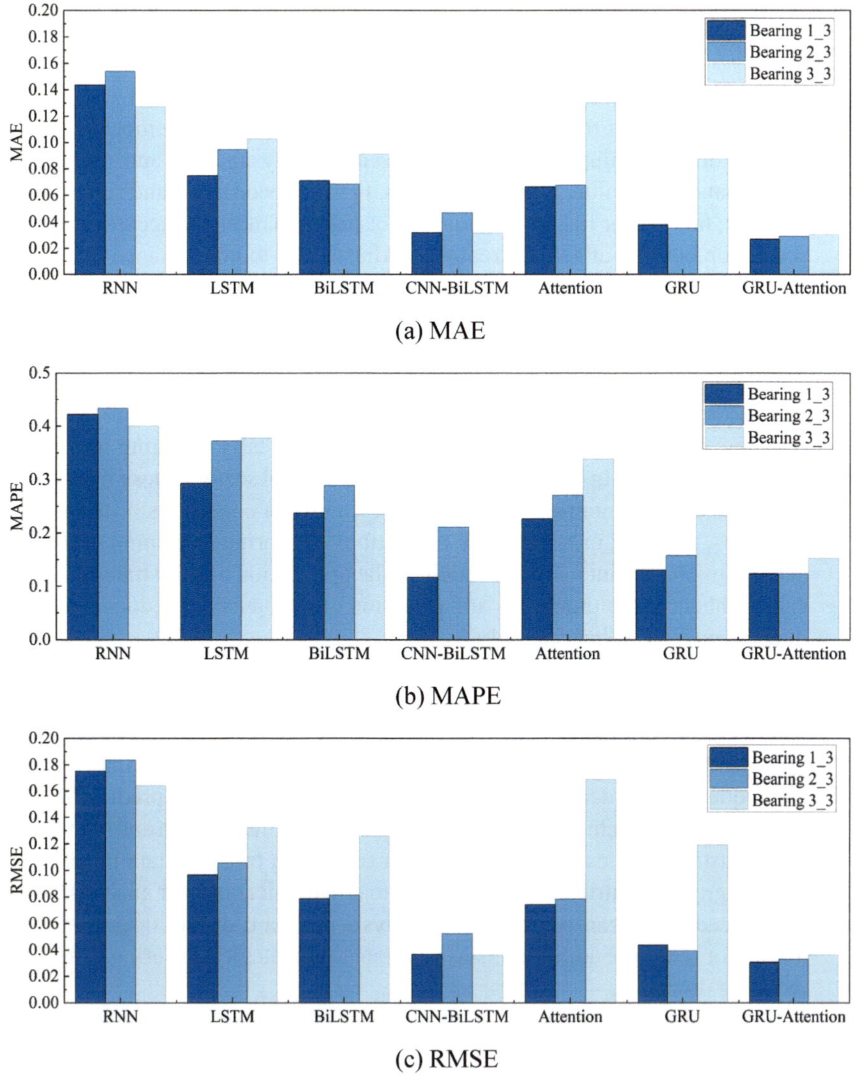

Fig. 4.10 Model performance comparison of all metrics

4.6 Conclusions

Bearings RUL predicting plays a crucial role in establishing predictive maintenance plans for IES. This not only allows machinery to operate for extended periods at minimal maintenance costs but also reduces the incidence of failures and enhances the safety and reliability of the equipment. Here are some potential applications of deep learning methods for bearing RUL prediction in the PHM of intelligent transportation systems:

(1) The application of deep learning combined with sensor fusion technology. By integrating data from various sensors (such as vibration, temperature, and sound), deep learning models can more accurately analyze and predict the performance and failure times of bearings. In high-speed trains and autonomous vehicles, this sensor fusion technology not only enhances the accuracy of fault prediction but also allows for real-time adjustments to maintenance schedules, thereby optimizing operational efficiency and safety. The implementation of this technology enables transportation systems to maintain high operational efficiency and safety standards in complex environments.
(2) The development of adaptive prediction and maintenance strategies. Adaptive prediction models automatically adjust their forecasting algorithms based on real-time data, making them well-suited for dynamically changing transportation environments. In the aviation and high-speed rail sectors, these models can adjust prediction parameters based on varying flight conditions or track environments, enabling more accurate predictions of bearing remaining useful life and optimizing maintenance schedules. The application of this strategy significantly enhances the timeliness and precision of maintenance, reducing the risk of accidents due to bearing failures.
(3) Integration of Internet of Things (IoT) and machine learning methods into smart monitoring systems. IoT technology enables the real-time collection of data from critical components such as bearings, including vibration and temperature, processed on the spot via edge computing. Utilizing machine learning techniques such as deep learning, these data can be analyzed to predict the RUL of bearings. This technology is particularly crucial in the aviation and high-speed rail industries. For example, the aviation sector can transmit data in real-time back to ground control centers via satellite communications for analysis, while high-speed trains can perform data analysis on-board during operation. This allows for real-time adjustments to maintenance strategies, ensuring safety and reducing downtime.
(4) Blockchain-based transparent and traceable maintenance record systems. The introduction of blockchain technology can significantly enhance the data integrity, security, and transparency of bearing RUL prediction based on deep learning. By creating an immutable and fully transparent data storage and sharing environment, blockchain not only increases the credibility of data and facilitates cross-organizational sharing but also supports real-time and dynamic data updates. This enhancement boosts the accuracy and adaptability of deep learning models, ultimately improving the maintenance efficiency and operational safety of the entire transportation system. For instance, maintenance records of autonomous vehicles can be used to bolster user trust. For the aviation industry, this technology can be employed to streamline safety reviews and accelerate maintenance response times.

References

1. Cai C, Qu ZJ, Ma LQ, Yu LF, Liu WB, Ren CG (2024) Spatial-temporal traffic flow prediction with fusion graph convolution network and enhanced gated recurrent units. IEEE Access 12:56477–56491. https://doi.org/10.1109/access.2024.3349690
2. Cheng YW, Hu K, Wu J, Zhu HP, Shao XY (2021) A convolutional neural network based degradation indicator construction and health prognosis using bidirectional long short-term memory network for rolling bearings. Adv Eng Inf 48:119682. https://doi.org/10.1016/j.aei.2021.101247
3. Gao CX, Lv K, Wu T, Si JK, Hu YH (2019) Method for determining starting point of rolling bearing life prediction based on linear regression. Electronics 8(9):923. https://doi.org/10.3390/electronics8090923
4. Gazizulin D, Klein R, Bortman J (2017) Towards efficient spall generation simulation in rolling element bearing. Fatigue Fract Eng Mater Struct 40(9):1389–1405. https://doi.org/10.1111/ffe.12580
5. Graves A, Schmidhuber J (2005) Framewise phoneme classification with bidirectional LSTM and other neural network architectures. Neural Netw 18(5–6):602–610. https://doi.org/10.1016/j.neunet.2005.06.042
6. Guo L, Li NP, Jia F, Lei YG, Lin J (2017) A recurrent neural network based health indicator for remaining useful life prediction of bearings. Neurocomputing 240:98–109. https://doi.org/10.1016/j.neucom.2017.02.045
7. Guo K, Ma J, Wu JD, Xiong X (2024) Adaptive feature fusion and disturbance correction for accurate remaining useful life prediction of rolling bearings. Eng Appl Artif Intell 138:109433. https://doi.org/10.1016/j.engappai.2024.109433
8. Guo RX, Wang YG (2021) Remaining useful life prognostics for the rolling bearing based on a hybrid data-driven method. Proc Inst Mech Eng Part I: J Syst Control Eng 235(4):517–531. https://doi.org/10.1177/0959651820948284
9. Karatug Ç, Ceylan BO, Arslanoglu Y (2025) A hybrid predictive maintenance approach for ship machinery systems: a case of main engine bearings. J Mar Eng Technol 24(1):12–21. https://doi.org/10.1080/20464177.2024.2393484
10. Li F, Liu SH, Wang TH, Liu RR (2024) Optimal planning for integrated electricity and heat systems using CNN-BiLSTM-Attention network forecasts. Energy 309:133042. https://doi.org/10.1016/j.energy.2024.133042
11. Li HF, Wei JL, Li SH, Liu YQ, Gu XH, Liu ZC, Yang SP (2023) Fatigue life prediction of high-speed train bearings based on the generalized linear cumulative damage theory. Fatigue Fract Eng Mater Struct 46(6):2112–2120. https://doi.org/10.1111/ffe.13984
12. Liu ZL, Zuo MJ, Qin Y (2016) Remaining useful life prediction of rolling element bearings based on health state assessment. Proc Inst Mech Eng Part C: J Mech Eng Sci 230(2):314–330. https://doi.org/10.1177/0954406215590167
13. Merainani B, Laddada S, Bechhoefer E, Chikh MAA, Benazzouz D (2022) An integrated methodology for estimating the remaining useful life of high-speed wind turbine shaft bearings with limited samples. Renew Energy 182:1141–1151. https://doi.org/10.1016/j.renene.2021.10.062
14. Nectoux P, Gouriveau R, Medjaher K, Ramasso E, Chebel-Morello B, Zerhouni N, Varnier C (2012) PRONOSTIA: an experimental platform for bearings accelerated degradation tests. In: IEEE international conference on prognostics and health management, PHM'12. IEEE Catalog Number: CPF12PHM-CDR, pp 1–8
15. Nie L, Zhang LF, Xu SY, Cai WT, Yang HM (2022) Remaining useful life prediction for rolling bearings based on similarity feature fusion and convolutional neural network. J Braz Soc Mech Sci Eng 44(8):328. https://doi.org/10.1007/s40430-022-03638-0
16. Pei H, Si XS, Hu CH, Zheng JF, Li TM, Zhang JX, Pang ZN (2021) An adaptive prognostics method for fusing CDBN and diffusion process: application to bearing data. Neurocomputing 421:303–315. https://doi.org/10.1016/j.neucom.2020.09.021

17. Peng P, Xu Y, Shi SQ, Fang ZY (2024) Application of non-Gaussian feature enhancement extraction in gated recurrent neural network for fault detection in batch production processes. Expert Syst Appl 237:121348. https://doi.org/10.1016/j.eswa.2023.121348
18. Qin CY, Qin DL, Jiang QX, Zhu BZ (2024) Forecasting carbon price with attention mechanism and bidirectional long short-term memory network. Energy 299:131410. https://doi.org/10.1016/j.energy.2024.131410
19. Que ZJ, Jin XH, Xu ZG (2021) Remaining useful life prediction for bearings based on a gated recurrent unit. IEEE Trans Instrum Meas 70:3511411. https://doi.org/10.1109/tim.2021.3054025
20. Shen F, Yan RQ (2022) A new intermediate-domain SVM-based transfer model for rolling bearing RUL prediction. IEEE-ASME Trans Mechatron 27(3):1357–1369. https://doi.org/10.1109/tmech.2021.3094986
21. Tajiani B, Vatn J, Naseri M (2024) Optimizing the maintenance threshold in presence of shocks: a numerical framework for systems with non-monotonic degradation. Reliab Eng Syst Saf 245:110039. https://doi.org/10.1016/j.ress.2024.110039
22. Wang FT, Liu XF, Deng G, Yu XG, Li HK, Han QK (2019) Remaining Life prediction method for rolling bearing based on the long short-term memory network. Neural Process Lett 50(3):2437–2454. https://doi.org/10.1007/s11063-019-10016-w
23. Wang FT, Wang B, Dun BS, Chen XT, Yan DW, Zhu H (2016) Remaining life prediction of rolling bearing based on PCA and improved logistic regression model. J Vibroeng 18(8):5192–5203. https://doi.org/10.21595/jve.2016.17449
24. Wang XL, Jiang B, Ding SX, Lu NY, Li Y (2022) Extended relevance vector machine-based remaining useful life prediction for DC-link capacitor in high-speed train. IEEE Trans Cybern 52(9):9746–9755. https://doi.org/10.1109/tcyb.2020.3035796
25. Xia M, Li T, Shu TX, Wan JF, de Silva CW, Wang ZR (2019) A two-stage approach for the remaining useful life prediction of bearings using deep neural networks. IEEE Trans Ind Inf 15(6):3703–3711. https://doi.org/10.1109/tii.2018.2868687
26. Xu ZQ, Zhang YJ, Miao Q (2024) An attention-based multi-scale temporal convolutional network for remaining useful life prediction. Reliab Eng Syst Saf 250:110288. https://doi.org/10.1016/j.ress.2024.110288
27. Yasin M, Sarigül M, Avci M (2024) Logarithmic learning differential convolutional neural network. Neural Netw 172:106114. https://doi.org/10.1016/j.neunet.2024.106114
28. Yu YZ, Cao Y, Wang G, Pang YJ, Lang LY (2023) Optical diffractive convolutional neural networks implemented in an all-optical way. Sensors 23(12):5749. https://doi.org/10.3390/s23125749

Chapter 5
Optimization Based Prediction for IGBT Remaining Useful Life

5.1 Introduction

Permanent Magnet Synchronous Motor (PMSM) is widely used in the field of robotics [34]. IGBT is utilized for converting DC power into three-phase AC power of variable frequency and voltage to drive the PMSM [26]. IGBT can improve the working efficiency and control flexibility of PMSMs, but also enhances their reliability, promoting the widespread application of this technology in various industries, e.g., robots. Due to the complexity of the robot's working environment, there are extreme situations such as sudden start and stop, long-term high-load work, etc. As one of the most vulnerable components in the converter, the IGBT is easily damaged, which in turn causes current distortion, and PMSM control failure, reducing motion accuracy [31].

Fault-tolerant control is a key technology in robot systems, which aims to ensure that robots can still effectively perform tasks and maintain their performance when encountering faults or abnormal situations [19]. Robots may face various unexpected situations when performing tasks, such as sensor failure or motor failure, and fault-tolerant control can monitor the system status in real-time, quickly identify faults and take appropriate remedial measures. In the event of partial system failure, the robot can continue to complete the predetermined goals by dynamically adjusting the control strategy or changing the task execution method. This not only improves production efficiency but also reduces economic losses caused by equipment failure [11], especially in application scenarios in high-risk or complex environments. To achieve fault-tolerant control of the robot, it is necessary to know the current state and the RUL of the IGBT device and perform early intervention. The current RUL prediction method of the IGBT could be divided into the physical methods, and data-driven methods [16].

The physical methods could consider the structure, material, and mechanics of the IGBT device, and establish a physical state degradation model to estimate the RUL. Lai et al. carried out a multi-physics simulation of the welding-type IGBT

H. Liu et al., *Prognostics and Health Management for Intelligent Electromechanical Systems*, https://doi.org/10.1007/978-981-96-7218-9_5

in the wind turbine, and proposed a RUL prediction model considering the electric frequency influence [14]. Liao et al. established failure-physics mathematical degradation models for high-speed train IGBT from aspects of the load-damage relation, task-load relation, and interdependence relation [18].

The data-driven methods rely on the high non-linear fitting capacity of the intelligent algorithms, and generate prediction results via learning the data dependency automatedly. He et al. applied a Back-Propagation neural network, RF, ELM to predict the RUL [7]. Haque et al. utilized an auxiliary particle filter for RUL prediction with reducing effect of random noise [6]. Li et al. adopted the LSTM model for RUL prediction of IGBT, showing better performance than the classical methods with strong early prognosis capacity [17].

In this chapter, the optimization method will be utilized to improve IGBT RUL prediction performance and establish a fault-tolerant control diagram for a front-wheel steering mobile robot accordingly.

5.2 Data Acquisition

In this study, the IGBT Accelerated Aging Data Set from NASA is utilized for experiments [25]. The dataset used the thermal cycling method to achieve accelerated aging of the IGBT devices. The parameters, including collector-emitter current, collector-emitter voltage, gate-emitter voltage, package temperature, etc. are monitored. According to previous studies, the collector-emitter voltage V_{CE} could reflect the degradation process of the IGBT device [10]. In this study, the first 6 devices are used as the training dataset, and the 7th device is adopted as the testing dataset. The raw data are shown in Fig. 5.1.

5.3 Predictive Model for IGBT Remaining Useful Life Based on Particle Swarm Optimization

5.3.1 *Health Indicator Based on Particle Swarm Optimization*

The health indicator is used to reflect the degradation performance of the device [27]. A good health indicator should have high monotonicity, prognosability, and trendability. Assuming time is T, health index is H, the expressions of the monotonicity, prognosability, and trendability are shown as follows [2, 12, 29]:

$$monotonicity(H) = \frac{1}{M}\sum_{m=1}^{M}\left(1 - \frac{6\sum_{n=1}^{N_m} d_{n,m}^2}{N_m\left(N_m^2 - 1\right)}\right) \tag{5.1}$$

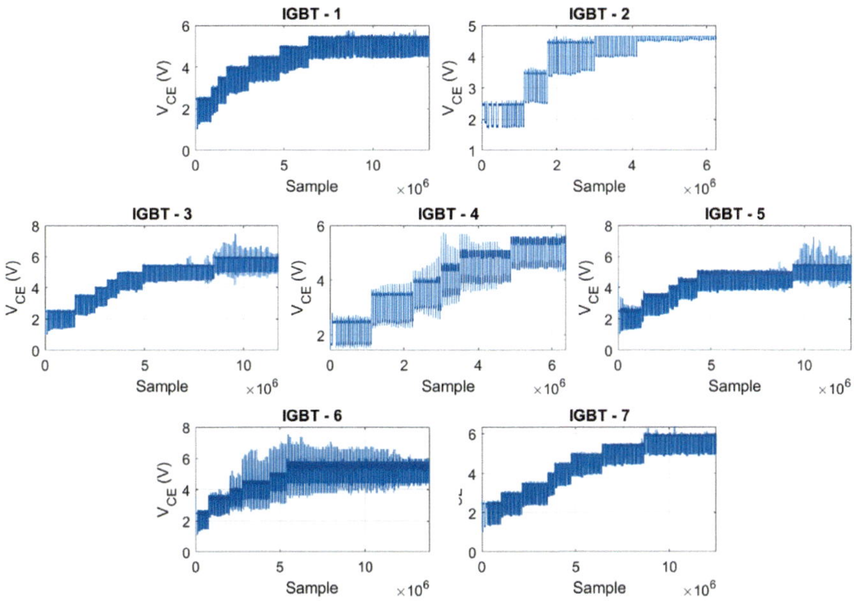

Fig. 5.1 Raw data of the IGBT V_{CE} parameter

where $d_n = R(T_n) - R(H_n)$, R is the rank function, the number of tested devices is M, the length of the corresponding test time of mth device is N_m. The monotonicity index is to calculate the Spearman's rank correlation between the time and health index [29].

$$prognosability(H) = \exp\left(-\frac{\text{std}_{m=1}^{M}\left(H_{N_m,m}\right)}{\text{mean}_{m=1}^{M}\left|H_{1,m} - H_{N_m,m}\right|}\right) \tag{5.2}$$

where $\text{std}_{m=1}^{M}$ and $\text{mean}_{m=1}^{M}$ is to calculate the standard deviation and average value of the variable along all tested devices. The prognosability index could measure the diversity of the health index at failure [2].

$$trendability(H) = \min_{i,j=1}^{M} corr\left(H_i, H_j\right) \tag{5.3}$$

where *corr* is a correlation function. The trendability index is set to quantify the similarity of the health index between different devices [12].

The commonly used features for IGBT RUL prediction contain Rooted Mean Sqrt (RMS), standard deviation, kurtosis, skewness, and Crest Factor (CF), etc. Taking the 7th IGBT device as an example, the features are shown in Fig. 5.2. The monotonicity, prognosability, and trendability of the features are shown in Table 5.1.

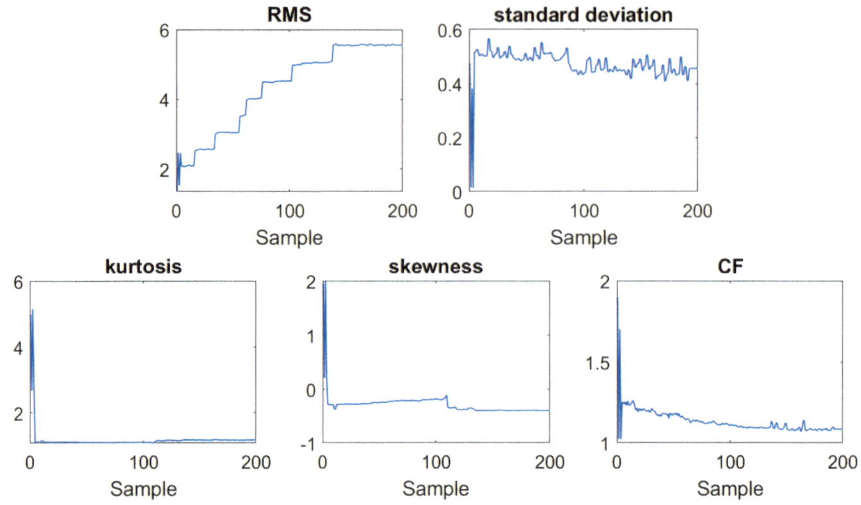

Fig. 5.2 Features of the IGBT V_{CE} parameter

Table 5.1 Quantitative evaluation of feature metrics for IGBT RUL prediction

Features	Monotonicity	Prognosability	Trendability
RMS	0.719	0.875	0.903
Standard deviation	0.131	0.219	0.148
Kurtosis	0.078	0.716	0.575
Skewness	0.060	0.371	0.078
CF	0.134	0.342	0.105

To better reflect the true degradation process of the IGBT, it is necessary to produce composite health indicators based on multiple features. In this book, the linear weight is used to build the composite health indicator, so as to improve the interpretability of the health indicator.

The particle swarm optimization (PSO) is used to optimize the following objective function:

$$\frac{1}{3}\left[monotonicity(H) + prognosability(H) + trendability(H)\right] \tag{5.4}$$

where the objective function could estimate the comprehensive performance of the monotonicity, prognosability, and trendability. The three functions in Eq. (5.4) all range from 0 to 1. Therefore, they are assigned with the same weights to construct the objective function which also ranges from 0 to 1.

The PSO is an optimization method based on swarm intelligence [13]. The core idea of the algorithm is to find the optimal solution to the problem by simulating the interaction of a group of particles. The PSO first randomly initializes particles

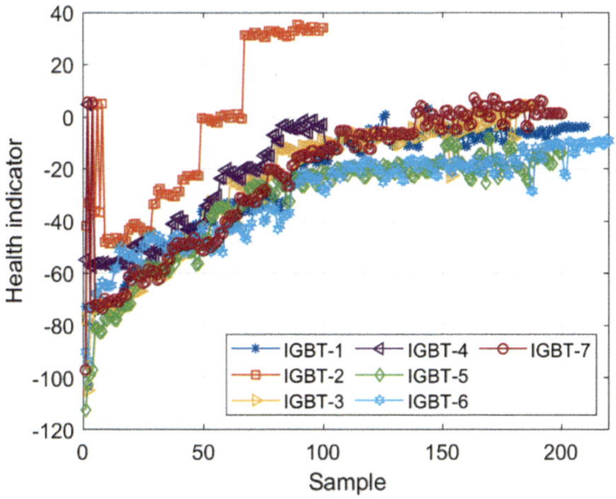

Fig. 5.3 Health indicator of the IGBT based on the PSO

in a predefined search space, each particle represents a potential solution and has two properties: position and velocity. In each iteration, the algorithm evaluates the fitness of each particle according to the objective function and updates the state of each particle until the set stop condition is reached, such as the maximum number of iterations or the fitness reaches a predetermined threshold. PSO has the advantages of a simple structure and fewer parameters. It can converge quickly and is widely used in various optimization problems, including engineering design [8], machine learning [20], etc. After optimization, the health indicator series of the devices are shown in Fig. 5.3. It can be observed that the health indicator generally shows an increasing trend, showing the effectiveness of the proposed method.

5.3.2 RUL Prediction Based on the Similarity

In this study, the method based on the similarity is applied to predict the RUL. The similarity-based method assumes the devices have identical degradation dynamics. Thus, the trend of the health index of the tested device could be estimated by finding the most matched curve from the degradation template pool. The DTW method is widely used to quantify the similarity [4, 22]. The core of the DTW method is to find the minimal distance between two time series x_i and y_j via setting pair-wise mapping function $\pi_{i,j}$. The minimization expression is shown as follows [32]:

$$\min_{\pi}\left[\sum_{(i,j)\in\pi} d\left(x_i, x_j\right)^2\right]^{0.5} \tag{5.5}$$

where π obeys $\pi_1 = (1, 1)$, $\pi_K = (K, K)$, and the π increases monotonously.

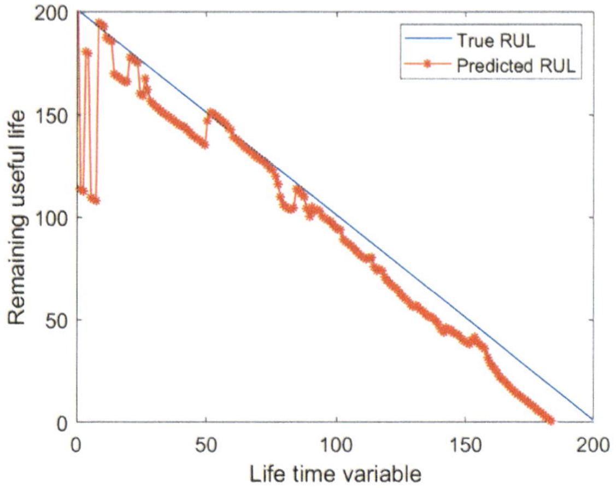

Fig. 5.4 Predicted RUL of the similarity-based method with PSO

Using the DTW method, we can find the closest curves of the health indicator from the template and calculate the RUL accordingly. The results are shown in Fig. 5.4. It can be observed that the predicted RUL fluctuates at the beginning. With the amount of the monitored data increases, the method could better find the curve template and produce stable prediction results.

5.4 Predictive Model for IGBT Remaining Useful Life Based on Bat Optimization

Similar to the PSO in Sect. 5.3, the Bat Algorithm (BA) is used to construct the health indicator and generate RUL prediction results in this section. The BA is a heuristic optimization algorithm based on the echolocation behavior of bats [20], which aims to simulate the navigation mechanism of bats when searching for prey. The algorithm first randomly generates a group of bats in the search space, each bat represents a potential solution and has properties such as position speed and emission frequency. The core process of the algorithm includes fitness evaluation, position and speed update, and random search until the preset stop condition is met. The Bat Algorithm has many advantages, including powerful global optimization ability, good adaptability, and fast convergence speed. In addition, since the algorithm has fewer parameters, parameter adjustment is relatively simple. BA has been widely used in many fields, including data mining [9], image processing [1], etc., especially in solving complex multi-objective optimization problems, demonstrating its potential as an efficient optimization tool.

The results based on the BA are shown in Figs. 5.5 and 5.6.

In the rest of the section, the emphasis is the application of RUL prediction in four-wheel mobile robot fault-tolerant control.

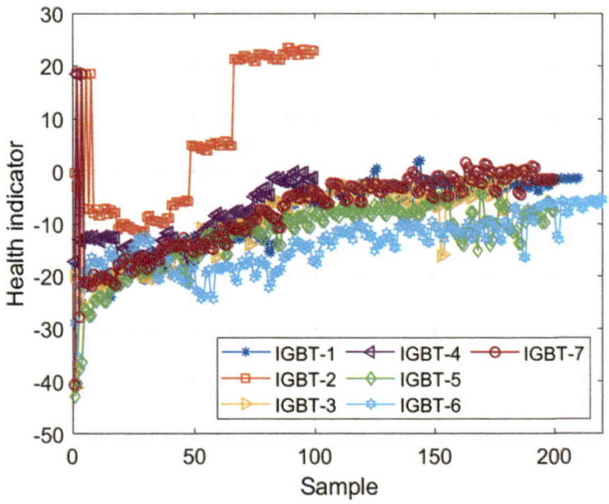

Fig. 5.5 Health indicator of the IGBT based on the BA

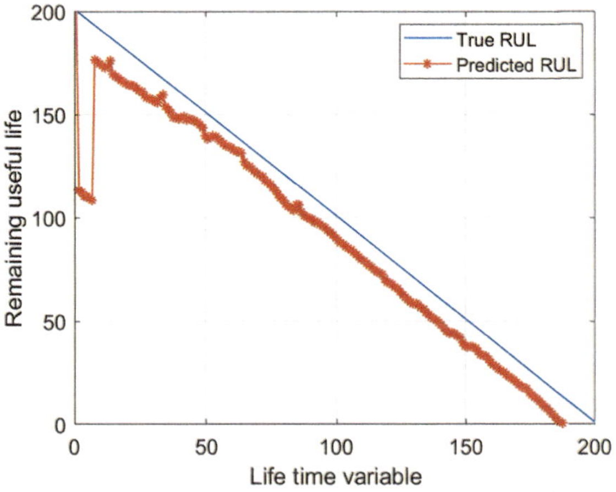

Fig. 5.6 Predicted RUL of the similarity-based method with BA

5.5 Model Performance Comparison

In this section, the performance of the PSO and BA on the IGBT RUL prediction is compared with three performance indexes, including MAE, RMSE and Pearson's correlation (P). The definitions of the three indexes are shown as follows:

$$MAE = \frac{\sum_{i=1}^{N} |x_i - \hat{x}_i|}{N} \tag{5.6}$$

$$RMSE = \sqrt{\frac{\sum_{i=1}^{N} (x_i - \hat{x}_i)^2}{N}} \tag{5.7}$$

Table 5.2 Model performance comparison

Index	PSO	BA
MAE	12.8640	14.3144
RMSE	18.2336	22.6476
P	0.9774	0.9549

$$P = \frac{\sum X\hat{X} - \frac{\sum X \sum \hat{X}}{N}}{\sqrt{N\sum X^2 - \left(\sum X\right)^2}\sqrt{N\sum \hat{X}^2 - \left(\sum \hat{X}\right)^2}} \tag{5.8}$$

where x and $\hat{x}$ are real and predicted RUL respectively, and N is the number of the samples.

The MAE, RMSE, and P are commonly used evaluation indicators, each with different focuses. MAE measures the average bias between the predicted and the actual values. It is simple and insensitive to outliers. RMSE amplifies the impact of large errors and is more suitable for scenarios that focus on large errors. P measures the linear correlation between the predicted and actual values, reflecting the consistency of the trend rather than the error size. In general, MAE and RMSE are used to evaluate errors, and P is used to evaluate correlation. The combination of indicators could reflect the comprehensive performance of the models.

From Table 5.2, it can be seen that the prediction results of the PSO (particle swarm optimization) model on the research dataset are better than those of the BA (bat algorithm) model. However, the scope of application of this conclusion is limited to the current dataset and may not be universal. This is consistent with the core idea of the "No Free Lunch Theorem", that is, in all possible problem spaces, no algorithm can always outperform other algorithms. The performance of any optimization algorithm depends on the specific characteristics and conditions of the problem.

In actual research, over-reliance on a single algorithm may lead to limitations. Different algorithms have their own unique applicable scenarios and performance advantages, so it is particularly important to build a diverse model library. The model library can include not only classic algorithms (such as PSO, BA, GA, etc.), but also emerging optimization methods. This diversity allows researchers to flexibly select or combine algorithms according to the characteristics of the problem, thereby improving the reliability and applicability of the results.

In addition, another direction worthy of attention is the exploration of multi-model fusion (Ensemble Learning). By combining the prediction results of multiple models, the limitations of a single model can be reduced, and the overall performance can be improved. For example, integrating the advantages of PSO and BA to form a hybrid optimization algorithm may show better results in more scenarios.

In summary, in order to fully tap the potential of optimization algorithms in practical applications, researchers need to adopt a comprehensive and flexible strategy, continuously expand and improve the model library, and conduct targeted experiments based on the characteristics of the data set. This will not only help improve the practicality of the algorithm but also provide more diverse and accurate tool support for solving complex problems.

5.6 Application in Front-Wheel Steering Mobile Robot Fault-Tolerant Control

5.6.1 Front-Wheel Steering Mobile Robot System

In this study, a front-wheel steering mobile robot is used [3]. The dynamic bicycle model, as a simplified Ackerman front-wheel steering robot model, is shown in Fig. 5.7. The front-wheel steering robot model is a simplified kinematic model, whose main components include the front wheel (responsible for steering), the rear wheel (providing driving force), the center of mass (used to describe the position), the wheelbase (the distance between the front and rear wheels), and the control input (the front wheel steering angle and the rear wheel driving speed). The model is characterized by being able to accurately describe the robot's motion behavior in a two-dimensional plane, suitable for path tracking and navigation in low-speed scenarios, and has the advantages of being simple and easy to implement, but it ignores the dynamic effects of tire sideways and slippage, and has limited accuracy at high speeds or in complex terrains.

The mathematical expression of the dynamic bicycle model with faults is given as follows [24]:

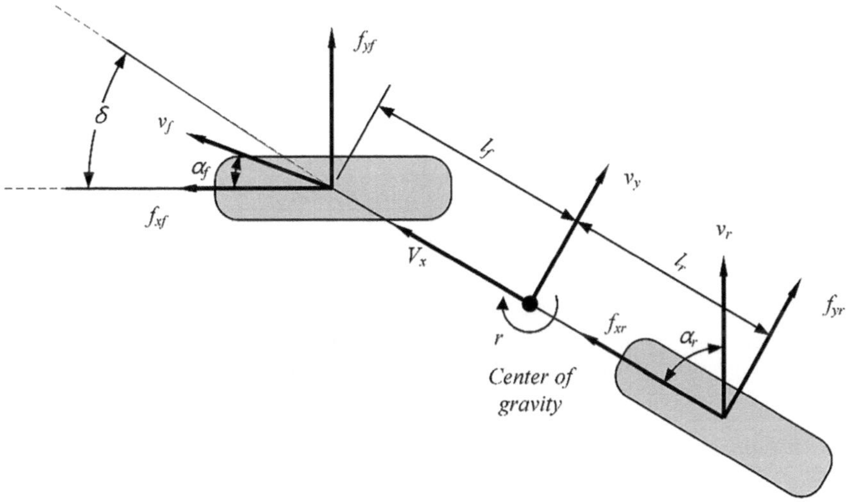

Fig. 5.7 Dynamic bicycle model of the front-wheel steering mobile robot

$$\begin{bmatrix} \dot{v}_y \\ \dot{r} \end{bmatrix} = \begin{bmatrix} \frac{-(c_f+c_r)}{mv_x} & \frac{l_r c_r - l_f c_f}{mv_x} - v_x \\ \frac{l_r c_r - l_f c_f}{I_z v_x} & \frac{-(l_f^2 c_f + l_r^2 c_r)}{I_z v_x} \end{bmatrix} \begin{bmatrix} v_y \\ r \end{bmatrix} + \begin{bmatrix} \frac{c_f}{m} \\ \frac{l_f c_f}{m} \end{bmatrix} \delta + \begin{bmatrix} d_1 \\ d_2 \end{bmatrix} \tag{5.9}$$

where v_y and r are the lateral speed and rotation speed respectively, α_f and α_r are the ship angles of the front and rear wheels respectively, c_f and c_r are the coefficient to define the linear correlation between ship angles and x-axis force respectively, l_f and l_r are the distance between front wheel, rear wheel and the center of gravity respectively, I_z is the yew inertia, m is the mass, δ is the front wheel angle, d is the external disturbance caused by faults.

In this study, the longitudinal speed v_x is set as the constant value. The front wheel angle δ is controlled to adjust the lateral speed v_y and rotation speed r, so as to trace the per-defined path. As shown in Fig. 5.8, the trace error is defined as follows [24]:

$$\begin{aligned} \begin{bmatrix} \dot{e}_{cg} \\ \ddot{e}_{cg} \\ \dot{\theta}_e \\ \ddot{\theta}_e \end{bmatrix} &= \begin{bmatrix} 0 & 1 & 0 & 0 \\ 0 & \frac{-(c_f+c_r)}{mv_x} & \frac{c_f+c_r}{m} & \frac{l_r c_r - l_f c_f}{mv_x} \\ 0 & 0 & 0 & 1 \\ 0 & \frac{l_r c_r - l_f c_f}{I_z v_x} & \frac{l_f c_f - l_r c_r}{I_z} & \frac{-(l_f^2 c_f + l_r^2 c_r)}{I_z v_x} \end{bmatrix} \begin{bmatrix} e_{cg} \\ \dot{e}_{cg} \\ \theta_e \\ \dot{\theta}_e \end{bmatrix} \\ &+ \begin{bmatrix} 0 \\ \frac{c_f}{m} \\ 0 \\ \frac{l_f c_f}{I_z} \end{bmatrix} \delta + \begin{bmatrix} 0 \\ \frac{l_r c_r - l_f c_f}{mv_x} - v_x \\ 0 \\ \frac{-(l_f^2 c_f + l_r^2 c_r)}{I_z v_x} \end{bmatrix} \varphi + \begin{bmatrix} 0 \\ d_1 \\ 0 \\ d_2 \end{bmatrix} \end{aligned} \tag{5.10}$$

where $\dot{e}_{cg} = v_y + v_x \sin\theta_e$ and $\theta_e = \theta - \theta_p$ are the position and angle error respectively, φ is the changing rate of the path tangential angle.

Let $\varepsilon_1 = \left[e_{cg}, \theta_e\right]^T, \varepsilon_2 = \left[\dot{e}_{cg}, \dot{\theta}_e\right]^T$, the error equation Eq. (5.11) can be expressed as follows:

$$\begin{aligned} \dot{\varepsilon}_1 &= \varepsilon_2 \\ \dot{\varepsilon}_2 &= \begin{bmatrix} 0 & \frac{c_f+c_r}{m} \\ 0 & \frac{l_f c_f - l_r c_r}{I_z} \end{bmatrix} \varepsilon_1 + \begin{bmatrix} \frac{-(c_f+c_r)}{mv_x} & \frac{l_r c_r - l_f c_f}{mv_x} \\ \frac{l_r c_r - l_f c_f}{I_z v_x} & \frac{-(l_f^2 c_f + l_r^2 c_r)}{I_z v_x} \end{bmatrix} \varepsilon_2 \\ &+ \begin{bmatrix} \frac{l_r c_r - l_f c_f}{mv_x} - v_x \\ \frac{-(l_f^2 c_f + l_r^2 c_r)}{I_z v_x} \end{bmatrix} \varphi + \begin{bmatrix} d_1 \\ d_2 \end{bmatrix} + \begin{bmatrix} \frac{c_f}{m} \\ \frac{l_f c_f}{I_z} \end{bmatrix} \delta \\ &= F_1 \cdot \varepsilon_1 + F_2 \cdot \varepsilon_2 + F_3 \cdot \varphi + D + H \cdot \delta \end{aligned} \tag{5.11}$$

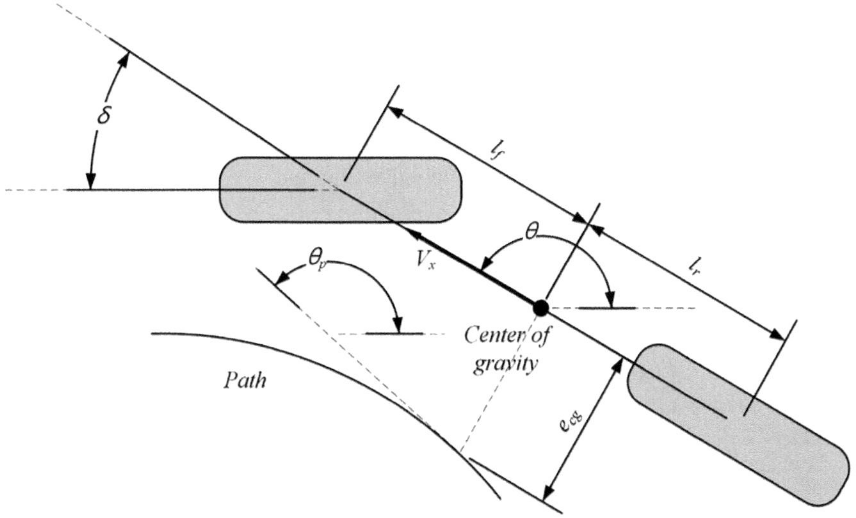

Fig. 5.8 Dynamic bicycle model for path tracking error analysis

5.6.2 Control Design

Sliding Mode Control (SMC) is a nonlinear control method based on the theory of variable structure systems [23]. It is mainly used to solve the control problems of complex systems with strong nonlinearity, uncertainty, or external disturbances. Its basic principle is to guide the system state to the sliding surface and keep it moving along the specified trajectory on the sliding surface by designing the sliding surface and switching control law, so as to achieve robust control of the target system. The main advantage of sliding mode control lies in its strong robustness to model uncertainty and external disturbances because once entering the sliding surface, the movement of the system is only defined by the sliding surface and has nothing to do with the specific model parameters. In addition, sliding mode control also has dynamic decoupling characteristics, which can achieve high control accuracy and fast dynamic response. However, due to the inherent discreteness of the switching control law, sliding mode control is prone to control chattering, which may lead to system instability or actuator loss [15]. In general, sliding mode control has become one of the important methods in the field of modern nonlinear control due to its simple theory, strong adaptability, and wide range of engineering application scenarios.

Nonsingular Fast Terminal Sliding Mode (NFTSM) is an improved sliding mode control method that aims to improve the dynamic response speed and convergence performance of the control system while avoiding the singularity problem that may occur in traditional terminal sliding mode control [33]. The core principle of NFTSM is to design a nonlinear sliding surface so that the system state can converge at a faster speed when approaching the equilibrium point and avoid the situation where the denominator tends to zero in the entire state space to ensure the effectiveness of the controller. Compared with traditional sliding mode control, NFTSM also has the advantage of a less chatter phenomenon, which is mainly due to its nonlinear smoothing design, which can improve the smoothness of the control input and thus reduce the burden on the actuator. NFTSM is widely used in fields that require high precision and fast dynamic response, such as spacecraft attitude control [5], drone trajectory tracking [30], industrial robot control [28], etc.

Considering surface of the NFTSM:

$$\Sigma = \varepsilon_1 + \Lambda_1 \text{sign}^{\Gamma_1} \varepsilon_1 + \Lambda_2 \text{sign}^{\Gamma_2} \varepsilon_2 \tag{5.12}$$

where the sliding surface Σ has the same dimension with the state parameter ε, $\Lambda_1 = diag(\lambda_{11}, \lambda_{12}, \ldots, \lambda_{1n})$, $\Lambda_2 = diag(\lambda_{21}, \lambda_{22}, \ldots, \lambda_{2n})$, $\Gamma_1 = diag(\gamma_{11}, \gamma_{12}, \ldots, \gamma_{1n})$, $\Gamma_2 = diag(\gamma_{21}, \gamma_{22}, \ldots, \gamma_{2n})$, $\lambda_1 > 0$, $\lambda_2 > 0$, $\gamma_1 > \gamma_2$, $1 < \gamma_2 < 2$.

The $\text{sign}^{\Gamma_1} \varepsilon_1$ and $\text{sign}^{\Gamma_2} \varepsilon_2$ are defined as follow:

$$\Lambda_1 \text{sign}^{\Gamma_1} \varepsilon_1 = \left[\text{sign}(\varepsilon_{11}) |\varepsilon_{11}|^{\Gamma_{11}}, \text{sign}(\varepsilon_{12}) |\varepsilon_{12}|^{\Gamma_{12}}, \ldots, \text{sign}(\varepsilon_{1n}) |\varepsilon_{1n}|^{\Gamma_{1n}} \right] \tag{5.13}$$

$$\Lambda_2 \text{sign}^{\Gamma_2} \varepsilon_2 = \left[\text{sign}(\varepsilon_{21}) |\varepsilon_{21}|^{\Gamma_{21}}, \text{sign}(\varepsilon_{22}) |\varepsilon_{22}|^{\Gamma_{22}}, \ldots, \text{sign}(\varepsilon_{2n}) |\varepsilon_{2n}|^{\Gamma_{2n}} \right] \tag{5.14}$$

Then, the control law becomes:

$$\delta = -H^{-1} \left[\begin{array}{l} M_2 \Sigma + (\rho + M_1) \dfrac{\Sigma}{\|\Sigma\|} + F_1 \cdot \varepsilon_1 + F_2 \cdot \varepsilon_2 + F_3 \cdot \varphi \\ + \Lambda_2^{-1} \Gamma_2^{-1} \left(I_n + \Lambda_1 \Gamma_1 diag\left(|\varepsilon_1|^{\Gamma_1 - I_n} \right) \text{sign}^{2I_n - \Gamma_2} (\varepsilon_2) \right) \end{array} \right] \tag{5.15}$$

where $M_1 > 0$, $M_2 > 0$, I_n is the unit matrix of n dimension, the coefficient ρ is a value to quantify the effects of the fault to the controller.

Theorem: the proposed control law Eq. (5.15) with the sliding surface Eq. (5.12) could make system Eq. (5.11) converges globally, considering coefficient ρ is the upper bound of the disturbance D: $\|D\| < \rho$.

Proof: The Lyapunov function candidate is selected as:

$$V = \frac{1}{2} \Sigma^T \Sigma \tag{5.16}$$

Its derivatives along time is:

$$\begin{aligned} \dot{V} &= \Sigma^T \dot{\Sigma} \\ &= \Sigma^T \left(\begin{array}{l} \dot{\varepsilon}_1 + \Lambda_1 \Gamma_1 diag\left(|\varepsilon_1|^{\Gamma_1 - I_n} \right) \text{sign}^{\Gamma_1} \dot{\varepsilon}_1 \\ + \Lambda_2 \Gamma_2 diag\left(|\varepsilon_2|^{\Gamma_2 - I_n} \right) \text{sign}^{\Gamma_2} \dot{\varepsilon}_2 \end{array} \right) \end{aligned}$$

$$= \Sigma^T \begin{pmatrix} \dot{\varepsilon}_1 + \Lambda_1 \Gamma_1 diag\left(|\varepsilon_1|^{\Gamma_1 - I_n}\right)\text{sign}^{\Gamma_1} \dot{\varepsilon}_1 \\ +\Lambda_2 \Gamma_2 diag\left(|\dot{\varepsilon}_1|^{\Gamma_2 - I_n}\right)\text{sign}^{\Gamma_2}(F_1 \cdot \varepsilon_1 + F_2 \cdot \varepsilon_2 + F_3 \cdot \varphi + D + H \cdot \delta) \end{pmatrix} \tag{5.17}$$

According to Eq. (5.15), Eq. (5.17) can be rewritten as follows:

$$\begin{aligned} \dot{V} &= -\Sigma^T \Lambda_2 \Gamma_2 diag\left(|\dot{\varepsilon}_1|^{\Gamma_2 - I_n}\right)\left((\rho + M_1)\frac{\Sigma}{\|\Sigma\|} + M_2 \Sigma - D\right) \\ &= -\Lambda_2 \Gamma_2 diag\left(|\dot{\varepsilon}_1|^{\Gamma_2 - I_n}\right)\left(\left(\rho\frac{\Sigma^T \Sigma}{\|\Sigma\|} - D\Sigma^T\right) + M_1 \frac{\Sigma \Sigma^T}{\|\Sigma\|} + M_2 \Sigma \Sigma^T\right) \end{aligned} \tag{5.18}$$

It is easily to know the $M_1 \frac{\Sigma\Sigma^T}{\|\Sigma\|} + M_2 \Sigma\Sigma^T > 0$ as $M_1 > 0, M_2 > 0$. Therefore, if $\rho \frac{\Sigma^T\Sigma}{\|\Sigma\|} - D\Sigma^T > 0$, then $\dot{V} < 0$:

$$\begin{aligned} &\rho \frac{\Sigma^T \Sigma}{\|\Sigma\|} - D\Sigma^T > 0 \\ &\rho \Sigma^T \Sigma - \Sigma^T D\left(\Sigma^T \Sigma\right)^{0.5} > 0 \\ &\rho - \frac{\left(\Sigma_1 D_1 + \Sigma_2 D_2 + \cdots + \Sigma_n D_n\right)}{\sqrt{\Sigma_1^2 + \Sigma_2^2 + \cdots + \Sigma_n^2}} > 0 \end{aligned} \tag{5.19}$$

Considering $\|D\| = \sqrt{D_1^2 + D_2^2 + \ldots + D_n^2} < \rho$, Eq. (5.19) yields:

$$\begin{aligned} &\sqrt{D_1^2 + D_2^2 + \cdots + D_n^2} - \frac{\Sigma_1 D_1 + \Sigma_2 D_2 + \cdots + \Sigma_n D_n}{\sqrt{\Sigma_1^2 + \Sigma_2^2 + \cdots + \Sigma_n^2}} > 0 \\ &\sqrt{D_1^2 + D_2^2 + \cdots + D_n^2}\sqrt{\Sigma_1^2 + \Sigma_2^2 + \cdots + \Sigma_n^2} > \Sigma_1 D_1 + \Sigma_2 D_2 + \cdots + \Sigma_n D_n \end{aligned} \tag{5.20}$$

According to Cauchy–Schwarz inequality, Eq. (5.20) holds. So, the proof is completed.

5.6.3 Simulation Results

The parameters are set as Table 5.3 to verify the effeteness of the proposed controller. In order to fully verify the effectiveness of the proposed controller, two representative cases can be used for evaluation. The first case is the ideal straight line-to-curve transformation. This ideal model allows us to test in a controlled environment. This process can understand the performance limit of the controller under ideal conditions, thus providing a basis for its parameter adjustment. The second case focuses on the actual application of real robots and faces more complex challenges. The experimental design of this case will examine the robustness and adaptability of the controller in a dynamic environment and evaluate its stability and response characteristics under uncertain conditions.

(1) *Simulation on the ideal curve*

In this section, a step signal of the curvature is used to verify the proposed control scheme. The curve is visualized in Fig. 5.9. The equation of the curve is presented in Eq. (5.18). The first part of the curve is a straight line, the second part is a circular with a radius of 40 m. The robot starts from the connection point of the two curves. Although the two curves are connected, the curvature is stepped from ∞ to 0.025. It is an ideal example the test whether the proposed controller can track the change.

$$Y = \begin{cases} 40 & X < 0 \\ \sqrt{1600 - X^2} & X \geq 0 \end{cases} \tag{5.21}$$

Setting the initial track error as $\varepsilon_1(0) = [0.2, 0.2]^T$, the error plot is shown in Fig. 5.10. As can be seen from Fig. 5.8, the system can converge effectively in about 5 s, and the longitudinal error e_{cg} and directional error θ_e can be effectively controlled.

The surface plot of the proposed NFTSM control is shown in Fig. 5.11. The figure also shows that the proposed control method can make the system state quickly approach the sliding plane and maintain it on the plane, but there is a serious chattering phenomenon.

The phase space plot of the proposed NFTSM control is shown in Fig. 5.12. In Fig. 5.12, it can also be seen that the phase space of the system can reach stable status,

Table 5.3 Simulation parameters for controller evaluation

Parameter	Value
Disturbance D based by the failure	$\left[300 \sin 2\pi t \ 20\pi \ \sin 3\pi t\right]^T$
Disturbance coefficient ρ	$\sqrt{300^2 + 20\pi^2}$
Time step	0.0001 s
c_f, c_r	150,000 N/rad, 150,000 N/rad
l_f, l_r	1 m, 1 m
m	1000 kg
v_x	1 m/s
I_z	1400 kg·m^2
Λ_1, Λ_2	$diag(1, 1), diag(1, 1)$
Γ_1, Γ_2	$diag(2, 2), diag(1.5, 1.5)$
M_1, M_2	2, 2

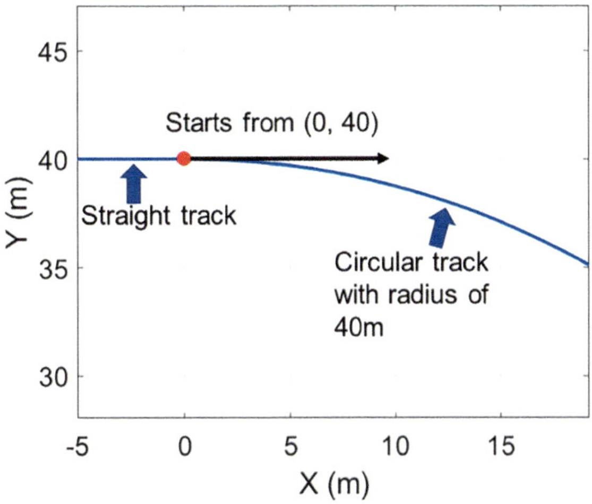

Fig. 5.9 XY trajectory with curved path for path tracking simulation

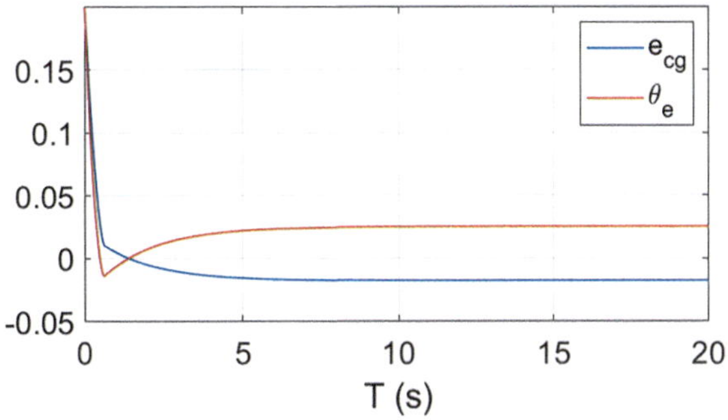

Fig. 5.10 Tracking error response of the NFTSM controller

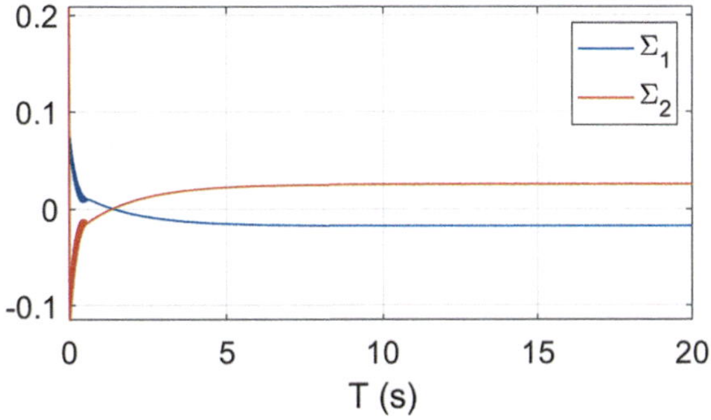

Fig. 5.11 Surface plot of sliding surface convergence under NFTSM control

followed by large-scale chattering, which will seriously affect the actual application capability of the model.

The surface plot of the proposed NFTSM control is shown in Fig. 5.13. As can be seen from Fig. 5.11, at the beginning of the control, the wheel steering angle δ is

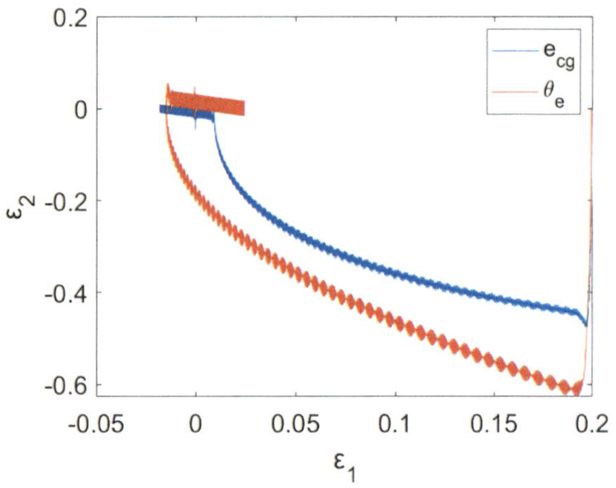

Fig. 5.12 Phase space evolution with NFTSM control

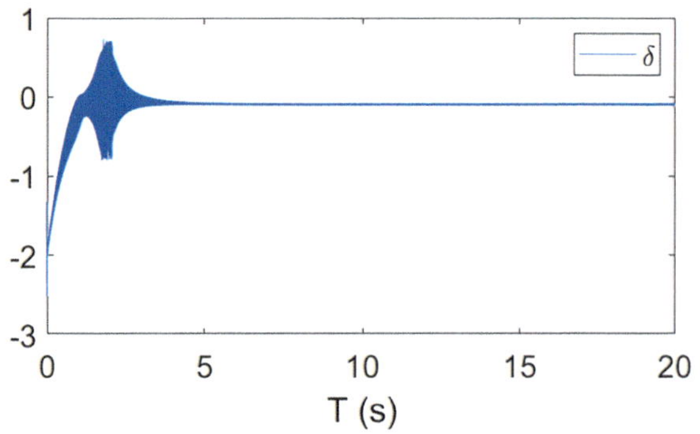

Fig. 5.13 Control signal plot of the NFTSM control

large, which then gradually decreases and reaches equilibrium after vibration. This is because the curve is circular and only the steering angle needs to be maintained to achieve tracking.

To alleviate the above chattering phenomenon, the following saturation sign function is used instead of the sign function in Eqs. (5.12) and (5.15):

$$\widetilde{sign}(x) = \begin{cases} 1 & x \geq \phi \\ x/\phi & \phi > x \geq -\phi \\ -1 & -\phi \end{cases} \tag{5.22}$$

where ϕ is set as 1 in this study.

Then, the control signal plot with the saturation sign function is shown in Fig. 5.14. As can be seen from Fig. 5.14, the chattering situation has been effectively alleviated and the system can approach the stable point in a relatively smooth manner.

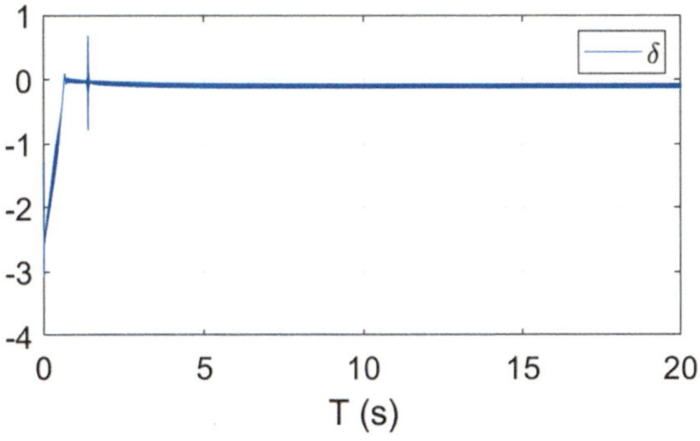

Fig. 5.14 Control signal plot of the NFTSM control with smoothed sign function

The tracking performance of the controller is shown in Fig. 5.15. It can be seen from the figure that the proposed method can well ensure that the robot moves along the predetermined track, which shows the effectiveness of the proposed method.

(2) *Simulation on the real robot trajectory*

Considering the real trajectory of the robot in Fig. 5.16. The data are collected from the Characterizing Visual Localization and Mapping Datasets (https://peringlab.org/lmdata/) [21].

Setting the initial track error as $\varepsilon_1(0) = [0.2, 0.2]^T$, the error plot, surface plot and phrase space plot are shown in Figs. 5.17, 5.18 and 5.19. It can be seen from the figure that our proposed method can converge effectively and control the tracking accuracy of the robot on complex paths.

The tracking performance of the controller on the real trajectory is shown in Fig. 5.20. It can be seen from the figure that the proposed method can well ensure that the robot moves along the predetermined track, although the real trajectory is

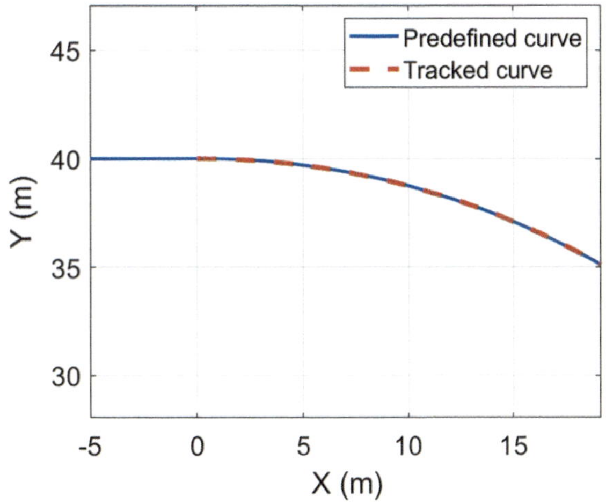

Fig. 5.15 Tracking performance of the NFTSM control

Fig. 5.16 Real trajectory of a robot

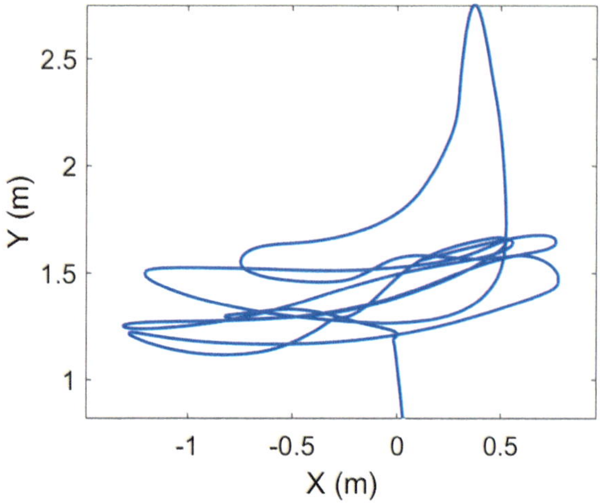

Fig. 5.17 Time-domain error response on complex path

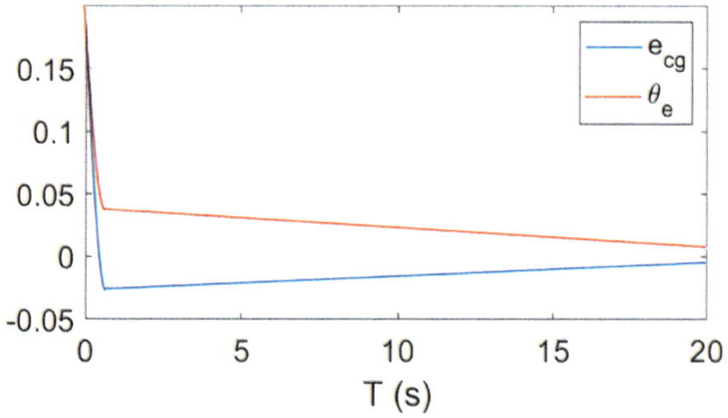

Fig. 5.18 Surface plot of tracking error under NFTSM control

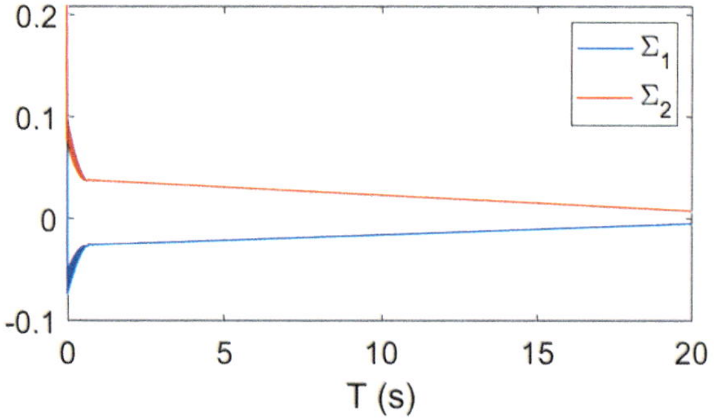

complex, and the initial error is large. The phenomenon shows the effectiveness of the proposed method.

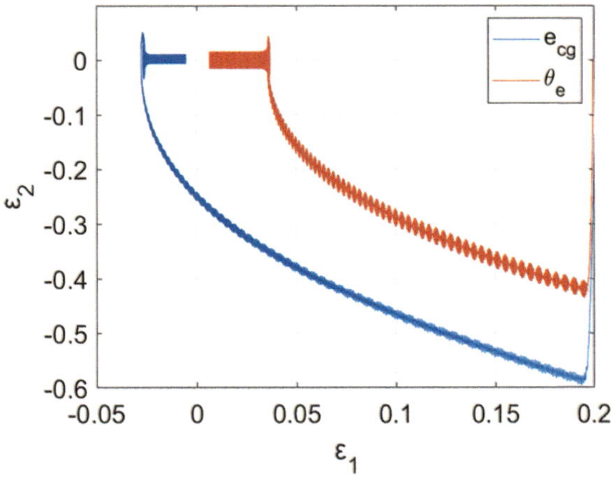

Fig. 5.19 Phase space trajectory of tracking errors

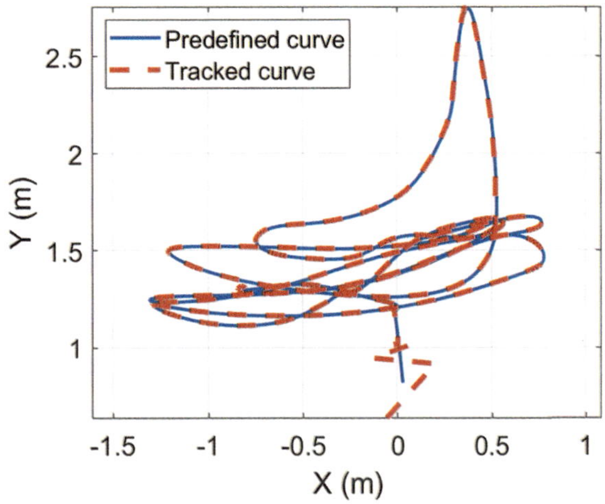

Fig. 5.20 Tracking performance of the NFTSM control on real trajectory

5.7 Conclusions

This study proposes a novel method for predicting the remaining life of IGBT faults, aiming to improve the reliability and safety of power electronic equipment. To this end, the template matching method is used in the study to accurately evaluate the remaining service life of the existing IGBT by comparing its operating characteristics with historical fault data. The advantage of this method is that the richer the historical data, the better the model performance. It can more effectively identify potential fault risks, thereby providing a scientific basis for equipment maintenance and replacement.

On this basis, the study also proposed a fault-tolerant control method for a four-wheeled robot based on the NFTSM (nonlinear fuzzy sliding mode control) theory.

This method ensures accurate trajectory tracking of the robot in a complex environment by analyzing the inherent disturbances that the robot may encounter during operation. Using the NFTSM control theory, a robust control system can be designed to cope with various uncertainties and disturbances and improve the adaptability of the robot in a dynamic environment.

In addition, the concept of fault parameter ρ is introduced in the study, which is used to evaluate the characteristics of fault disturbances. In the control model, the dynamic adjustment of fault parameters enables the control system to respond to the operating status of the equipment in real-time, ensuring that a stable control effect can be maintained when a fault occurs. By combining with the remaining life of the IGBT model, the study provides a data-driven evaluation framework for practical applications, which can achieve timely updates and monitoring of fault parameters in the actual operation of the equipment.

Overall, this study not only deepens the understanding of IGBT fault prediction and robot control technology in theory but also provides a practical method in practical applications, providing new ideas and a practical basis for fault-tolerant control and fault diagnosis in the field of power electronic equipment and intelligent robots.

References

1. Alihodzic A, Tuba M (2014) Improved bat algorithm applied to multilevel image thresholding. Sci World J 1:176718
2. Aziz NH, Catterson VM, Rowland SM, Bahadoorsingh S (2017) Analysis of partial discharge features as prognostic indicators of electrical treeing. IEEE Trans Dielectr Electr Insul 24(1):129–136
3. Bae B, Lee DH (2023) Design of a four-wheel steering mobile robot platform and adaptive steering control for manual operation. Electronics 12(16):3511
4. Guan XD, Huang C, Liu GH, Meng XL, Liu QS (2016) Mapping rice cropping systems in Vietnam using an NDVI-based time-series similarity measurement based on DTW distance. Remote Sens 8(1):19
5. Han ZG, Zhang K, Yang TS, Zhang MH (2016) Spacecraft fault-tolerant control using adaptive non-singular fast terminal sliding mode. IET Control Theory Appl 10(16):1991–1999
6. Haque MS, Choi S, Baek J (2018) Auxiliary particle filtering-based estimation of remaining useful life of IGBT. IEEE Trans Industr Electron 65(3):2693–2703. https://doi.org/10.1109/tie.2017.2740856
7. He CJ, Yu WZ, Zheng YZ, Gong WJ (2021) Machine learning based prognostics for predicting remaining useful life of IGBT–NASA IGBT accelerated ageing case study. In: 2021 IEEE 5th information technology, networking, electronic and automation control conference (ITNEC). IEEE, pp 1357–1361
8. He Q, Wang L (2007) An effective co-evolutionary particle swarm optimization for constrained engineering design problems. Eng Appl Artif Intell 20(1):89–99
9. Heraguemi KE, Kamel N, Drias H (2018) Multi-objective bat algorithm for mining numerical association rules. Int J Bio-Inspired Comput 11(4):239–248
10. Ismail A, Saidi L, Sayadi M, Benbouzid M (2020) A new data-driven approach for power IGBT remaining useful life estimation based on feature reduction technique and neural network. Electronics 9(10). https://doi.org/10.3390/electronics9101571
11. Kamel MA, Yu X, Zhang YM (2017) Fault-tolerant cooperative control design of multiple wheeled mobile robots. IEEE Trans Control Syst Technol 26(2):756–764

12. Keizers LS, Loendersloot R, Tinga T (2024) Towards a hybrid framework for prognostics with limited run-to-failure data. In: 8th European conference of the Prognostics and Health Management Society, PHME 2024. PHM Society, pp 844–855
13. Kennedy J, Eberhart R (1995) Particle swarm optimization. In: Proceedings of ICNN'95-international conference on neural networks. IEEE, pp 1942–1948
14. Lai W, Liu AB, Wang Z, Shangguan M, Li H, Chen MY, Yao R, Xiong Y (2023) Study on lifetime modeling of IGBT modules considering electric frequency influence mechanism. IEEE Trans Device Mater Reliab 23(3):395–403. https://doi.org/10.1109/tdmr.2023.3288144
15. Lee H, Utkin VI (2007) Chattering suppression methods in sliding mode control systems. Annu Rev Control 31(2):179–188
16. Li HQ, Zhang ZX, Li TM, Si XS (2024) A review on physics-informed data-driven remaining useful life prediction: challenges and opportunities. Mech Syst Signal Process 209. https://doi.org/10.1016/j.ymssp.2024.111120
17. Li WP, Wang BX, Liu JC, Zhang GG, Wang JH (2020) IGBT aging monitoring and remaining lifetime prediction based on long short-term memory (LSTM) networks. Microelectron Reliab 114. https://doi.org/10.1016/j.microrel.2020.113902
18. Liao J, Peng T, Xu YS, Gui G, Yang C, Yang CH, Gui WH (2024) Task-orientated probabilistic damage model with interdependent degradation behaviors for RUL prediction of traction converter systems. Reliab Eng Syst Saf 250. https://doi.org/10.1016/j.ress.2024.110302
19. Milecki A, Nowak P (2023) Review of fault-tolerant control systems used in robotic manipulators. Appl Sci 13(4):2675
20. Qolomany B, Maabreh M, Al-Fuqaha A, Gupta A, Benhaddou D (2017) Parameters optimization of deep learning models using particle swarm optimization. In: 2017 13th international wireless communications and mobile computing conference (IWCMC). IEEE, pp 1285–1290
21. Saeedi S, Carvalho EDC, Li W, Tzoumanikas D, Leutenegger S, Kelly PHJ, Davison AJ (2019) Characterizing visual localization and mapping datasets. In: Saeedi S, Carvalho EDC, Li W et al (eds) 2019 international conference on robotics and automation (ICRA), 20–24 May 2019, pp 6699–6705. https://doi.org/10.1109/ICRA.2019.8793528
22. Shen JY, Huang WP, Zhu DY, Liang J (2017) A novel similarity measure model for multivariate time series based on LMNN and DTW. Neural Process Lett 45:925–937
23. Shtessel Y, Edwards C, Fridman L, Levant A (2014) Sliding mode control and observation, vol 10. Springer
24. Snider JM (2009) Automatic steering methods for autonomous automobile path tracking. Robotics Institute, Pittsburgh, PA, Tech Rep CMU-RITR-09-08
25. Sonnenfeld G, Goebel K, Celaya Jose R (2008) An agile accelerated aging, characterization and scenario simulation system for gate controlled power transistors. In: 2008 IEEE AUTOTESTCON. IEEE, pp 208–215
26. Stabile A, Estima JO, Boccaletti C, Cardoso AJM (2014) Converter power loss analysis in a fault-tolerant permanent-magnet synchronous motor drive. IEEE Trans Industr Electron 62(3):1984–1996
27. Vignolles A, Chanthery E, Ribot P (2020) An overview on diagnosability and prognosability for system monitoring. In: European conference of the Prognostics and Health Management Society (PHM Europe)
28. Vo AT, Kang HJ (2018) An adaptive neural non-singular fast-terminal sliding-mode control for industrial robotic manipulators. Appl Sci 8(12):2562
29. Wang CS, Lu NY, Cheng YH, Jiang B (2018) Deep forest based multivariate classification for diagnostic health monitoring. In: 2018 Chinese control and decision conference (CCDC). IEEE, pp 6233–6238
30. Wang J, Zhao ZH, Zheng Y (2018) NFTSM-based fault tolerant control for quadrotor unmanned aerial vehicle with finite-time convergence. IFAC-PapersOnLine 51(24):441–446
31. Xu SQ, Chen XY, Liu F, Wang H, Chai Y, Zheng WX, Chen HT (2023) A novel adaptive SMO-based simultaneous diagnosis method for IGBT open-circuit faults and current sensor incipient faults of inverters in PMSM drives for electric vehicles. IEEE Trans Instrum Meas

32. Yadav M, Alam MA (2018) Dynamic time warping (DTW) algorithm in speech: a review. Int J Res Electron Comput Eng 6(1):524–528
33. Yang L, Yang JY (2010) Nonsingular fast terminal sliding-mode control for nonlinear dynamical systems. Int J Robust Nonlinear Control 21(16):1865–1879. https://doi.org/10.1002/rnc.1666
34. Yang Y, He Qiang FuCY, Liao SP, Tan P (2020) Efficiency improvement of permanent magnet synchronous motor for electric vehicles. Energy 213:118859

Chapter 6
Decomposition Based Prediction for MOSFET Remaining Useful Life

6.1 Introduction

There are two main types of motor driver technology for robots, IGBT and MOSFET. The IGBT has slower switching speeds and is often considered to be robust and cost-effective motor drive solutions for low-speed switching applications [17]. When the switching frequency exceeds 16 kH, MOSFET is usually used instead. MOSFET is more advantageous in motor drives than IGBT solutions [3], and their high speeds and other characteristics are well suited for complex motor drive solutions, such as industrial robots.

As the on-time of a MOSFET increases during use, the device power and performance will decline and the RUL will gradually shorten, which significantly impact the stable operation of the entire robotic system. Therefore, to improve system operation reliability and realize condition-based maintenance, accurate prediction of MOSFET RUL [2, 12] is required. Current MOSFET prediction methods can be categorized into two types: data-driven and physics of failure [7, 16, 19].

The physics of failure approach takes into full account the failure mechanisms of MOSFET, characterized by clear physical meaning and intuitive representation. Zhou et al. introduced two leakage current mechanisms to build a failure model based on the traditional MOSFET circuit model to analyze the failure mechanism [25]. Zhao et al. utilized the potential failure precursor (PFP) to construct a composite failure precursor to predict RUL and validated the results on MOSFET [24].

Data-driven methods offer the capability to predict the lifetime even when the failure mechanism is unclear, which has the advantages such as reduced computation time, enhanced accuracy, and accelerated design processes [6]. Chen et al. used a gray wolf optimized bi-directional long and short-term memory neural network to predict the RUL [2]. Dusmez et al. proposed a data-driven RUL estimation algorithm based on the linear approximation model [5]. Ren et al. combined a long short-term memory network with continuous variational mode decomposition to construct a MOSFET lifetime prediction model [12]. Li et al. utilized an echo state network

H. Liu et al., *Prognostics and Health Management for Intelligent Electromechanical Systems*, https://doi.org/10.1007/978-981-96-7218-9_6

(ESN) and developed particle filtering methods to update the model to predict the RUL of MOSFET [9].

In this chapter, the decomposition method will be utilized to improve the RUL prediction performance of MOSFET and to build a corresponding fault-tolerant control figure for wheeled mobile robots.

6.2 Data Acquisition

In this study, the experiments used NASA's MOSFET Thermal Overstress Aging Data Set [1]. This data set uses indirect thermal cycling for the accelerated aging of MOSFET devices. The monitored parameters include monitoring data for gate-source voltage, drain-source voltage, leakage current, and case temperature. Since the health state of semiconductor devices is difficult to be directly observed during operation, this chapter chooses to use single degradation eigenvector as the state observation value and reacts to the experimental device state by calculating the drain-source on-resistance [22]. In this study, the full-life data from five experiments of devices No. 8, No. 11, No. 12, and No. 14 are used as the training data set, and the full-life data from five experiments of device No. 9 are used as the test data set (Fig. 6.1).

6.3 Predictive Model for MOSFET Remaining Useful Life Based on Wavelet Packet Decomposition

6.3.1 *Feature Extraction Based on Wavelet Packet Decomposition*

MOSFET monitoring data uses signal processing techniques to extract relevant features from the time and frequency domains, which can effectively enhance the prediction capability. It can be argued that feature engineering plays a crucial role in determining the maximum potential performance of machine learning models. Assuming that the time series data is D, this chapter uses the WPD algorithm to decompose it into multiple subsequences.

WPD can decompose the acquired signal into multiple two-dimensional components, encompassing covariates and frequencies, thereby enabling feature extraction from the signal across various frequency bands and at different time points [20]. Compared with wavelet decomposition, WPD decomposes the high-frequency component and low-frequency component of the signal at the same time, which has a strong time–frequency localization decomposition capability [11, 23]. The feature vectors selectively focus on frequency bands in an adaptive manner, capturing the signal's full-frequency characteristics and enhancing the time–frequency resolution.

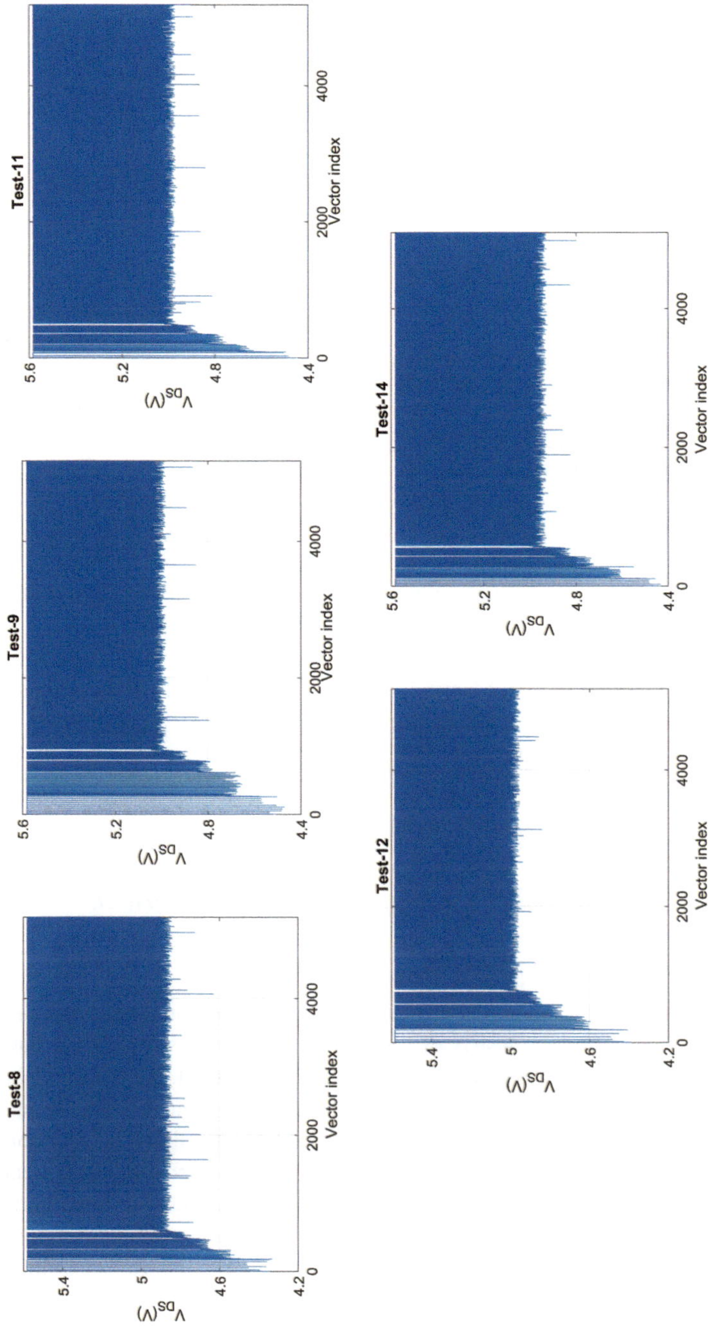

Fig. 6.1 Raw data of the MOSFET V_{DS} parameter

Table 6.1 Performance of the features

Features	RMS	Standard deviation	Kurtosis	Skewness	CF
MOSFET_8	5.1560	0.3447	1.3323	0.4128	1.0849
MOSFET_11	5.2538	0.2933	1.2454	0.2006	1.0643
MOSFET_12	5.2480	0.3050	1.2939	0.1420	1.0657
MOSFET_14	5.2213	0.3111	1.2762	0.2708	1.071
MOSFET_9	5.2839	0.2927	1.2793	– 0.0266	1.0577

The specific expression is shown below:

$$\mathrm{CWT}_f(a,b) = \int_{-\infty}^{+\infty} f(t)\psi^*((t-b)/a)/\sqrt{a}dt \tag{6.1}$$

where $f(t)$ is the signal, $\psi(t)$ is the mother wavelet function, $*$ is the complex conjugate, and a, b are the scale and translation coefficients, respectively.

The commonly used characteristics for MOSFET RUL prediction in line with IGBT include RMS, standard deviation, kurtosis, skewness, and CF. The first experimental characteristics of the experimental equipment selected for this study are shown in Table 6.1.

Using the WPD method, we take the first experimental data of experimental devices No. 8, No. 9, No. 11, No. 12, and No. 14 as an example, and the decomposition subsequence is shown in Fig. 6.2. Compared with wavelet decomposition, WPD is able to improve the time–frequency resolution of the signal, which has practical application value.

6.3.2 *The Theoretical Basis of Autoregressive Integrated Moving Average Model*

The full name of the ARIMA model is autoregressive integrated moving average model. The ARIMA model consists of three main components, which are autoregressive model, differential process and moving average model.

The basic idea of the ARIMA model is to use historical information about the data itself to predict the future. The model posits that labeled values oscillate around a prevailing temporal trend, which is shaped by the impact of past labels. The deviations from this trend are attributed to random events occurring over time, and it is acknowledged that the overall trend itself may not remain constant [13, 18].

In the ARIMA (p, q, d) model, p stands for the "Autoregressive" part: this part describes the lagged values of the observations used in the model (i.e., the values of the previous p periods). The starting point of the autoregressive model is to consider the observation as a linear combination of the p values that precede it. The exact

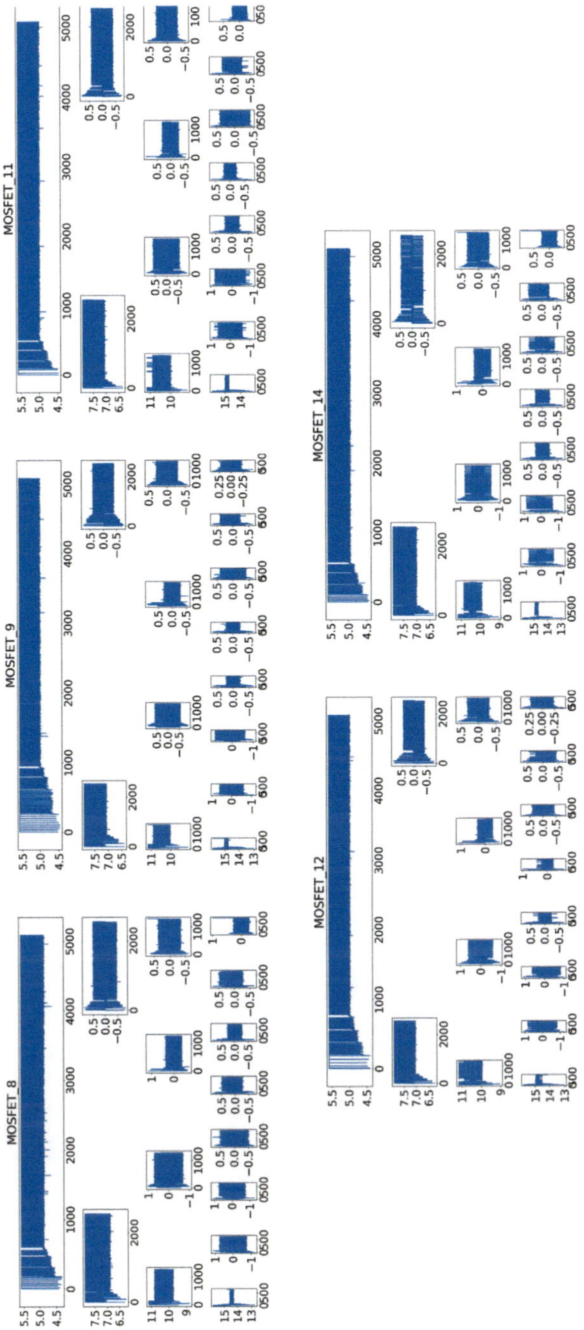

Fig. 6.2 Decomposition subsequences based on WPD

mathematical form is as follows:

$$AR : Y_t = c + \varphi_1 Y_{t-1} + \varphi_2 Y_{t-2} + \cdots + \varphi_p Y_{t-p} + \xi_t \tag{6.2}$$

where $\varphi_1, \varphi_2, \ldots, \varphi_p$ are model parameters, c is a constant, and ξ_t is white noise. The order of this equation determines the number of observations to which the model refers back.

q stands for the "Moving Average" part: this part describes the lagged value of the error term used in the model (i.e. the value of the previous q periods). The moving average model is a relationship between current values and past white noise. The exact mathematical form is as follows:

$$MA : Y_t = \mu + \epsilon_t + \theta_1 \epsilon_{t-1} + \theta_2 \epsilon_{t-2} + \cdots + \theta_q \epsilon_{t-q} \tag{6.3}$$

where $\theta_1, \theta_2, \ldots, \theta_q$ are model parameters, c is a constant, e_t is white noise in the current period, and $\epsilon_{t-1}, \epsilon_{t-2}, \ldots, \epsilon_{t-q}$ is white noise in the past. The order q of this equation determines the amount of white noise that the model looks back on.

d represents the order of the difference. The goal of differencing is to transform a non-smooth series into a smooth series. Suppose there is a time series y_t, define the lag operation B, which transforms an observation at one moment into an observation at a previous moment:

$$B_{y_t} = y_{t-1} \tag{6.4}$$

Extend this operator concept to lag by one time step:

$$B^n_{y_t} + y_{t-n} \tag{6.5}$$

The hysteresis operation is utilized to represent the first order difference and the n order difference. The first order difference can be viewed as the difference between neighboring labeled values, which can be expressed as:

$$\Delta y_t = y_t - y_{t-1} = y_t - B_{y_t} = (1 - B) y_t \tag{6.6}$$

The n-order difference is a subtraction of $n - 1$ labeled values apart:

$$\Delta^n y_t = y_t - y_{t-n} = y_t - B^n y_t = \left(1 - B^n\right) y_t \tag{6.7}$$

By analogy, the d-order difference can be expressed as (Fig. 6.3):

$$d\_order\_y = (1 - B)^d y_t \tag{6.8}$$

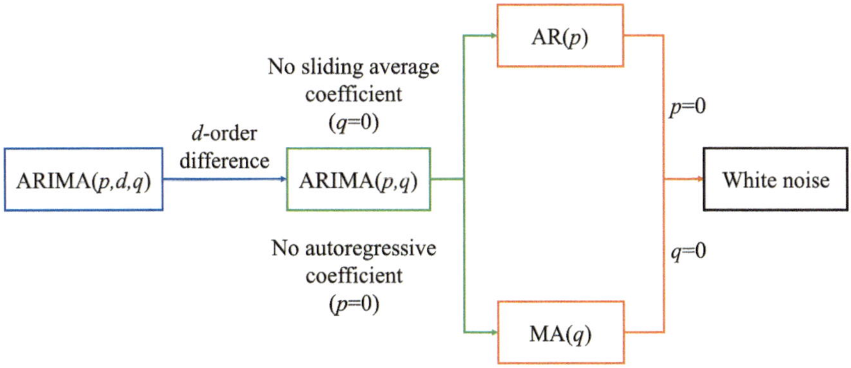

Fig. 6.3 ARIMA structure

6.3.3 *Experimental Result and Analysis*

In this section, ARIMA, Discrete Wavelet Transform (DWT)-ARIMA, EWT-ARIMA, and WPD-ARIMA models are developed to predict the remaining service life of MOSFET. The results of the four models are shown in Fig. 6.4. The WPD-ARIMA model has significantly superior performance compared to the other models. The results are analyzed as follows:

(1) Among all the models, ARIMA model has the worst performance. As can be seen from the figure, the prediction results of ARIMA have a large difference with the real value. The possible reason is that the time–frequency features in the original data are complex, and it is difficult for the simple ARIMA model to extract the hidden information effectively.

Fig. 6.4 ARIMA-based prediction results

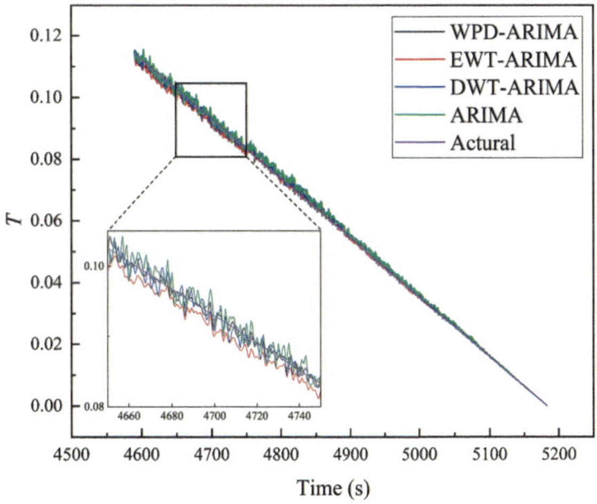

(2) The application of the DWT and EWT decomposition algorithms has resulted in an improved ARIMA model that is closer to the actual curve in terms of predicting the curve. This is mainly attributed to the fact that the two wavelet transforms can help the model determine the time-domain location of the different frequency components more accurately.
(3) WPD-ARIMA has the most significant performance with predictions closest to the true values. This is due to the fact that WPD can observe both the frequency and time axes, with better temporal resolution at high frequencies and better frequency resolution at low frequencies.

6.4 Predictive Model for MOSFET Remaining Useful Life Based on Complete Ensemble Empirical Mode Decomposition

6.4.1 Feature Extraction Based on Complete Ensemble Empirical Mode Decomposition

In this section, a method based on CEEMD is used to predict the MOSFET RUL. The CEEMD-based approach mitigates the issue of mode aliasing inherent in current methods, thereby enhancing the accuracy of RUL predictions in the forecasting domain. The CEEMD is based on the EEMD. It adds noise to the original signal, performs EEMD separately, and finally performs integrated averaging to obtain the decomposition results, which can effectively reduce white noise. The specific decomposition process is as follows:

Step 1: The ith signal $y^i(t)$ obtained by adding adaptive white noise to the original rainfall signal can be expressed as:

$$y^i(t) = y(t) + \varepsilon_0 \omega^i(t), \quad t = 1, 2, \dots, n \tag{6.9}$$

where n is the sample point of the data; t is the data period. $y(t)$ is the original signal. ε_0 is the standard deviation of the noise. $\omega^i(t)$ is the ith white noise signal.

Step 2: The EMD of $y^i(A)$ is performed until the first modal component $IMF_1(t)$ is obtained.

$$IMF_1(t) = \frac{1}{n}\sum_{i=1}^{n} IMF_1^i(t) \tag{6.10}$$

Calculate the first residual signal $r_1(t)$.

$$r_1(t) = y(t) - \overline{IMF_1(t)} \tag{6.11}$$

Step 3: The EMD algorithm is used to decompose the signal $r_1(t) + \varepsilon_1 E_1\left(\omega^i(t)\right)$ to obtain the second modal component $IMF_2(t)$.

$$IMF_2(t) = \frac{1}{n}\sum_{i=1}^{n} E_1\left(r_1(t) + \varepsilon_1 E_1\left(\omega^i(t)\right)\right) \tag{6.12}$$

For each of the remaining stages, i.e. $k = 2, \ldots, k$, the kth residual signal is calculated as follows:

$$r_k(t) = r_{k-1}(t) - IMF_k(t) \tag{6.13}$$

Repeat the procedure in step 3 to obtain the $k + 1$st modal component as follows:

$$IMF_{k+1}(t) = \frac{1}{n}\sum_{i=1}^{n} E_1\left(r_k(t) + \varepsilon_k E_k\left(\omega^i(t)\right)\right) \tag{6.14}$$

Steps 4 and 5: The above steps are repeated until the residual signal can no longer be decomposed, resulting in K modal components. The decomposed final residual signal is:

$$R(t) = y(t) - \sum_{i=1}^{K} IMF_k(t) \tag{6.15}$$

The original rainfall signal is therefore decomposed into:

$$y(t) = \sum_{i=1}^{K} IMF_k(t) + R(t) \tag{6.16}$$

Using the CEEMD method, we take the first experimental data of experimental devices No. 8, No. 9, No. 11, No. 12, and No. 14 as an example, and its decomposition subsequence is shown in Fig. 6.5 with many rounds of iterations, noise reduction and integration of signal components are realized, and the decomposition and feature extraction of the signal are completed.

6.4.2 The Theoretical Basis of Long Short-Term Memory Model

The LSTM network, a variant of RNN, is designed to address the gradient vanishing and exploding problems faced by traditional RNNs when dealing with long sequences. It incorporates a unique memory unit and gating mechanism to more

Fig. 6.5 Decomposition subsequences based on CEEMD

effectively capture and manage long-term dependencies within sequence data. The memory unit is the core of the LSTM network and is responsible for maintaining and updating the long-term dependency information throughout the sequence processing. The structure of the memory unit is relatively simple, consisting mainly of one or more neurons whose states are transmitted through time steps and updated only in a linear fashion [10, 14].

The forgetting gate serves the purpose of deciding what information from the memory unit should be discarded or retained at each time step. It is calculated by the following formula:

$$f_t = \sigma\left(W_f \cdot \left[h_{t-1}, x_t\right] + b_f\right) \tag{6.17}$$

where σ is the logical activation function, W_f and b_f are the weight matrix and bias term of the forgetting gate, respectively, and $\left[h_{t-1}, x_t\right]$ is the hidden state of the previous time step and the input of the current time step.

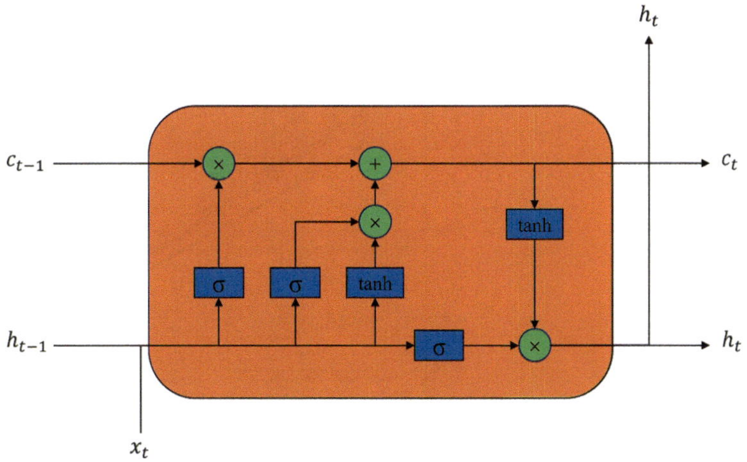

Fig. 6.6 LSTM structure

The input gate consists of two parts: a sigmoid layer that decides which values will be updated and a tanh layer that creates a new vector of candidate values that will be added to the state. The formula for the input gate is as follows:

$$i_t = \sigma\left(W_i \cdot \left[h_{t-1}, x_t\right] + b_i\right) \tag{6.18}$$

$$\tilde{C}_t = \tanh\left(W_C \cdot \left[h_{t-1}, x_t\right] + b_C\right) \tag{6.19}$$

where i_t is the output of the input gate, $\tilde{C}_t$ is the candidate memory cell state, W_i, W_C, and b_i, b_C are the correlation weight and bias, respectively.

The output gate is responsible for deciding which part of the memory cell state will be output to the hidden state, and is also calculated by the following equation:

$$o_t = \sigma\left(W_o \cdot \left[h_{t-1}, x_t\right] + b_o\right) \tag{6.20}$$

$$h_t = o_t * \tanh(C_t) \tag{6.21}$$

where o_t is the output of the sigmoid function of the output gate, C_t is the updated memory cell state, and h_t is the final hidden state output (Fig. 6.6).

6.4.3 Experimental Result and Analysis

In this section, LSTM, EMD-LSTM, VMD-LSTM, and CEEMD-LSTM models are developed to predict the remaining service life of MOSFET. The results of the four

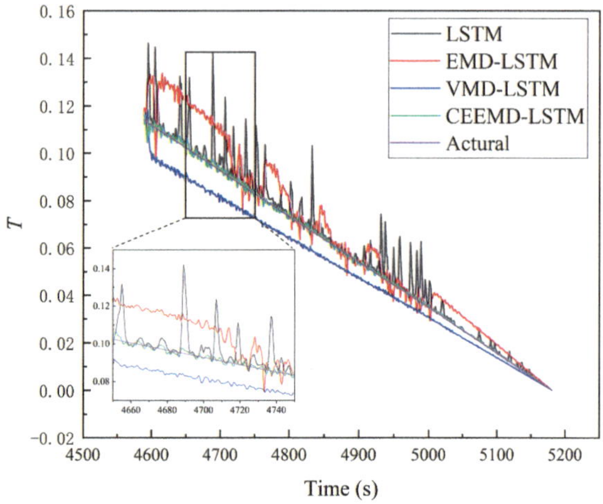

Fig. 6.7 LSTM-based forecast results

models are shown in Fig. 6.7. The CEEMD-LSTM model has the best performance. The results are analyzed as follows:

(1) Among all the models, LSTM has the worst performance. Compared to EMD-LSTM, VMD-LSTM and CEEMD-LSTM, LSTM prediction results are difficult to meet the desired requirements. The possible reason for this is that the lack of a data preprocessing step makes the model less capable of handling complex signals.
(2) The combination of EMD and LSTM results in improved model performance. The hybrid models EMD-LSTM and VMD-LSTM have higher prediction accuracy compared to single LSTM model. The possible reason is that EMD is suitable for analyzing nonlinear non-smooth signals with high signal-to-noise ratio and good time–frequency focusing. And VMD can be adjusted according to the signal characteristics to obtain the best decomposition.
(3) The use of CEEMD results in the highest model performance. This improvement is attributed to the fact that CEEMD adds positive and negative pairs of auxiliary white noise to the original signal, which effectively improves the decomposition efficiency and overcomes the problems of large reconstruction error and poor decomposition completeness of the original EMD.

6.5 Model Performance Comparison

After data preprocessing and dividing the training and test sets, MOSFET RUL prediction is performed using ARIMA algorithm and LSTM algorithm. In the time series prediction stage, the initial parameters of the LSTM model are drawn in a uniform distribution with mean 0 and variance $2/(m + n)$, where m is the number of inputs and n is the number of outputs. In addition, gradient optimization was performed using the Adam optimizer and trained to obtain the optimal model parameters.

Validate the model performance of RUL prediction based on MOSFET accelerated aging time series. The WPD-based model and the CEEMD-based model are compared. In addition, analyze the metrics of the prediction results.

The forecasting performance can be evaluated by three performance indices, including MAE, MAPE, and RMSE [4]. The equations of these indices are presented as follows [21]:

$$\begin{cases} MAE = \left(\sum_{t=1}^{N} \left|Y(t) - \hat{Y}(t)\right|\right)/N \\ MAPE = \left(\sum_{t=1}^{N} \left|\left(Y(t) - \hat{Y}(t)\right)/x(t)\right|\right)/N \\ RMSE = \sqrt{\left(\sum_{t=1}^{N} \left[Y(t) - \hat{Y}(t)\right]^2\right)/N} \end{cases} \tag{6.22}$$

where Y is the actual value, $\hat{Y}$ is the forecasting value [4, 8].

(1) The experimental results are shown in Table 6.2 and it can be seen that WPD-ARIMA predicts RUL best. Compared to other models, this method has lower MAE, MAPE and RMSE with less error fluctuation and can produce stable prediction results.
(2) The single ARIMA and LSTM models are deficient in RUL prediction for the MOSFET TOA dataset. The introduction of decomposition algorithms such as WPD and CEEMD, on the other hand, has led to an increase in the model's ability to handle complex non-stationary signals. This suggests that the decomposition algorithms are able to capture the time–frequency characteristics of the signals, allowing the model learning to focus on the key local content of the data.
(3) Through the use of WPD and CEEMD algorithms, the prediction accuracies of ARIMA and LSTM models were improved to different degrees. WPD led to

Table 6.2 Comparison of MOSFET RUL prediction performance

Model	MAE	MAPE	RMSE
ARIMA	6.0552	13.8588	8.8656
DWT-ARIMA	4.6789	12.3890	6.3951
EWT-ARIMA	2.6946	9.4232	3.7246
WPD-ARIMA	1.6217	6.7333	2.5257
LSTM	7.2789	14.5221	8.1094
EMD-LSTM	6.504	12.6067	7.375
VMD-LSTM	5.409	13.2039	6.0016
CEEMD-LSTM	2.592	10.1931	3.7566

an improvement of 51.42–73.22% in the performance of ARMIA metrics, and CEEMD led to an improvement of 29.81–64.39% in the performance of LSTM metrics. This shows that the decomposition algorithm is a key part of the data processing step, which is worth considering when building a machine learning framework (Fig. 6.8).

6.6 Applications in Wheeled Mobile Robot Fault-Tolerant Control

6.6.1 Fault-Tolerant Control

The kinematic analysis model of the designed wheeled mobile robot is established in Cartesian coordinate system. The wheeled mobile robot is a single rigid body with incomplete constraints and its kinematic equations can be expressed as:

$$\begin{bmatrix} \dot{x} \\ \dot{y} \\ \dot{\varphi} \end{bmatrix} = \begin{bmatrix} \cos\varphi & 0 \\ \sin\varphi & 0 \\ 0 & 1 \end{bmatrix} \begin{bmatrix} v \\ \omega \end{bmatrix} \tag{6.23}$$

where $\dot{x}$, $\dot{y}$ are the positions of the mobile robot in the plane; $\dot{\varphi}$ is the steering angle of the mobile robot; v is the value of the forward speed of the mobile robot; ω is the value of the steering speed of the mobile robot.

The relationship between the linear velocity of the two wheels and the forward and steering velocities of the robot:

$$\begin{bmatrix} v \\ \omega \end{bmatrix} = \begin{bmatrix} \frac{1}{2} & \frac{1}{2} \\ -\frac{1}{2l} & \frac{1}{2l} \end{bmatrix} \begin{bmatrix} v_L \\ v_R \end{bmatrix} \tag{6.24}$$

Finally, the system model of the wheeled mobile robot can be obtained as:

$$\begin{bmatrix} \dot{x} \\ \dot{y} \\ \dot{\varphi} \end{bmatrix} = \begin{bmatrix} \frac{r}{2}\cos\varphi & \frac{r}{2}\cos\varphi \\ \frac{r}{2}\sin\varphi & \frac{r}{2}\sin\varphi \\ -\frac{r}{2l} & -\frac{r}{2l} \end{bmatrix} \begin{bmatrix} k_m u_L \\ k_m u_R \end{bmatrix} \tag{6.25}$$

Failure of the MOSFET of a wheeled mobile robot manifests itself as a motor failure, which can affect the normal forward and steering motion of the robot. The fault based kinematic model of the wheeled mobile robot can be described as:

$$\dot{\chi} = f(\chi) + g(\chi)u(t) + F(t) \tag{6.26}$$

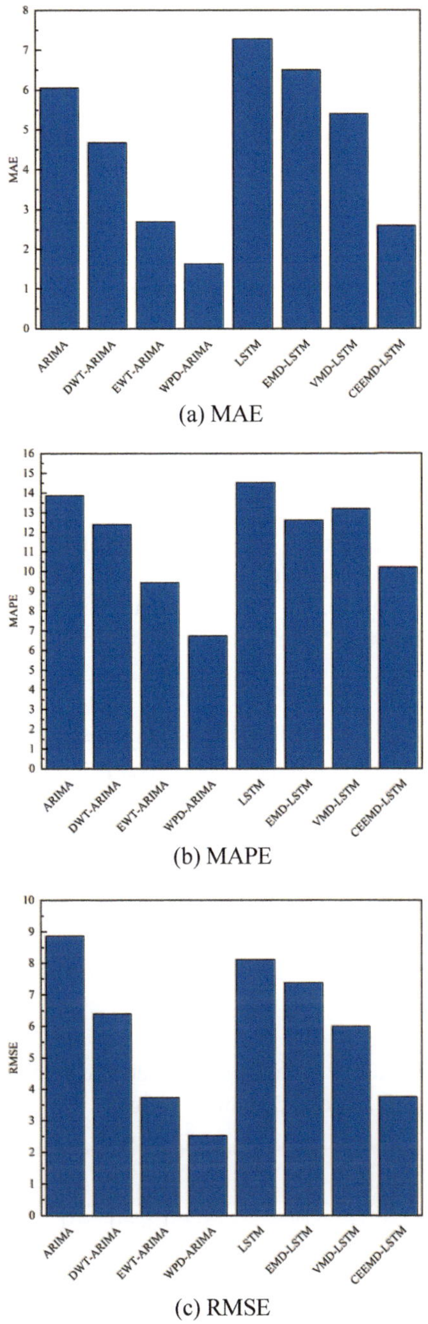

Fig. 6.8 Model performance comparison of all metrics

where $\chi = [\chi, y, \varphi]^T$ is the state quantity; $f(\chi) = [f_x, f_y, f_\varphi]^T$ is the fault information; $g(\chi)$ is the control input matrix; $F(t)$ is the fault information of the actuator.

Failure diagnostics FDI for wheeled mobile robots is:

$$\begin{cases} \hat{u} = g(\chi)^{-1}[\dot{\chi} - f(\chi)] \\ \hat{u} = u - [\hat{u} + f_{a_i}(t)] \end{cases} \tag{6.27}$$

where $\hat{u}$ is the observer model of the actuator; $\tilde{u}$ is the diagnostic error of the FDI.

At this point, $f_{a_i}(t)$ is described as:

$$f_{a_i}(k+1) = W_i\sigma\left(\sum_{j=1}^{m} V_{i,j}f_{a_i}(k-j+1) + B_i(k)\right) \tag{6.28}$$

where $B_i(k) = \sum_{j=1}^{n} V_{i,m+j}e_i(k-j+1)$.

The fault-tolerant control designed in this study consists of two components, i.e., a nonlinear dynamic inverse controller and an adaptive fault compensator. Based on the above, the kinematic equation of the wheeled mobile robot containing actuator faults is:

$$\begin{bmatrix} \dot{x} \\ \dot{y} \\ \dot{\varphi} \end{bmatrix} = \begin{bmatrix} f_x \\ f_y \\ f_\varphi \end{bmatrix} + g(\chi)\begin{bmatrix} k_m u_L \\ k_m u_R \end{bmatrix} \tag{6.29}$$

Let the parametric quantities of the wheeled mobile robot be x_d, y_d, and φ_d, then its nonlinear dynamic inverse controller can be designed as:

$$u_{DC} = g(\chi)^{-1}\left(\begin{bmatrix} \dot{x}_d \\ \dot{y}_d \\ \dot{\varphi}_d \end{bmatrix} - \begin{bmatrix} f_x \\ f_y \\ f_\varphi \end{bmatrix}\right) \tag{6.30}$$

where $\dot{x}_d$, $\dot{y}_d$ and $\dot{\varphi}_d$ are reference velocities.

$\dot{x}_d$, $\dot{y}_d$ and $\dot{\varphi}_d$ can be defined:

$$\begin{bmatrix} \dot{x}_d \\ \dot{y}_d \\ \dot{\varphi}_d \end{bmatrix} = \begin{bmatrix} k_1 & 0 & 0 \\ 0 & k_2 & 0 \\ 0 & 0 & k_3 \end{bmatrix}\begin{bmatrix} x_d - x \\ y_d - y \\ \varphi_d - \varphi \end{bmatrix} \tag{6.31}$$

The stability of $g(\chi)$ and its inverse matrix can be ensured by choosing appropriate controller parameters. Associating the above equations, the mathematical structure of the FTC can be obtained as:

$$\begin{bmatrix} \dot{x} \\ \dot{y} \\ \dot{\varphi} \end{bmatrix} = \begin{bmatrix} f_x \\ f_y \\ f_\varphi \end{bmatrix} + g(\chi) \begin{bmatrix} k_m u_L \\ k_m u_R \end{bmatrix} + F(t) - f_a(t) \tag{6.32}$$

where $f_a(t)$ can be used to compensate for actuator failures that have occurred.

Further, the AFTC controller for a wheeled mobile robot can be designed:

$$\begin{bmatrix} u_L \\ u_R \end{bmatrix} = k_m^{-1} g(\chi)^{-1} \left(\begin{bmatrix} \dot{x}_d \\ \dot{y}_d \\ \dot{\varphi}_d \end{bmatrix} - \begin{bmatrix} f_x \\ f_y \\ f_\varphi \end{bmatrix} + F(t) - f_a(t) \right) \tag{6.33}$$

6.6.2 Applications in Wheeled Mobile Robot

Fault-tolerant control of wheeled robots requires a combination of tracking and planning problems that allow vehicle states and control inputs to be considered in the cost function [15].

(a) **Wheeled Robot Model**: The wheeled robot model is employed for predicting robot behavior within the local plane, as it has demonstrated the ability to efficiently capture critical non-completeness constraints of the robot through computation.

The state vector and control vector are defined as follows:

$$\xi := \begin{bmatrix} x \\ y \\ \psi \\ u \end{bmatrix} = \begin{bmatrix} x \text{ position of center of mass} \\ y \text{ position of center of mass} \\ \text{yaw angle} \\ \text{longitudinal speed} \end{bmatrix} \tag{6.34}$$

$$\zeta := \begin{bmatrix} a_x \\ \delta_f \end{bmatrix} = \begin{bmatrix} \text{longitudinal acceleration} \\ \text{front tire steering angle} \end{bmatrix} \tag{6.35}$$

The nonlinear equation of state is defined as:

$$\dot{\xi} = \begin{bmatrix} u(t)\cos\left(\psi(t) + \beta\left(\delta_f\right)\right) \\ u(t)\sin\left(\psi(t) + \beta\left(\delta_f\right)\right) \\ u(t)\dfrac{\cos\left(\beta\left(\delta_f\right)\right)\tan\left(\delta_f\right)}{l_a + l_b} \\ a_x \end{bmatrix} \tag{6.36}$$

$$\beta(\delta_f) = \tan^{-1}\left(\frac{l_a}{l_a + l_b}\tan(\delta_f)\right) \tag{6.37}$$

The parameters l_a and l_b are the distances from the front and rear axes to the center of mass of the robot.

(b) **Cost function:** The cost function J is expressed as follows:

$$J = \omega_t t_f + \int_{t_0}^{t_f} \left(\omega_{\delta_f}\delta_f^2 + \omega_{a_x}a_x^2\right)dt + \int_{s_0}^{s_f} \omega_k k^2 dt \tag{6.38}$$

The first minimizes the total time to achieve the goal. The second term specifies the control work. The last term penalizes the curvature k along the path. The parameters ω_t, ω_{δ_f}, ω_{a_x} and ω_k are the corresponding weights of each term.

(c) **State and control boundaries:** The state and control boundaries are determined based on safety concerns and the mechanical constraints of the actuator. The state boundaries are extended as follows:

$$\begin{bmatrix} x_{\min} \\ y_{\min} \\ \psi_{\min} \\ u_{\min} \end{bmatrix} \le \begin{bmatrix} x \\ y \\ \psi \\ u \end{bmatrix} \le \begin{bmatrix} x_{\max} \\ y_{\max} \\ \psi_{\max} \\ u_{\max} \end{bmatrix} \tag{6.39}$$

The control range is set similarly as follows:

$$\begin{bmatrix} a_{x_{\min}} \\ \delta_{f_{\min}} \end{bmatrix} \le \begin{bmatrix} a_x \\ \delta_f \end{bmatrix} \le \begin{bmatrix} a_{x_{\max}} \\ k_{r_f}\delta_{f_{\max}} \end{bmatrix} \tag{6.40}$$

The degradation states of MOSFET during operation of wheeled mobile robots are categorized into normal loss, decline, and severe degradation periods. We incorporate the RUL prediction results into the control constraints of the wheeled mobile robot and reduce the loss by controlling the steering angle of the robot when the MOSFET is degraded to ensure the task execution. Where k_{r_x} denotes the front wheel steering angle control coefficient in the current state.

(d) **Terminal state constraints:** Constraints are imposed on the final state to ensure the robot attains the intended destination. They are expressed as follows:

$$\begin{bmatrix} x_f - \varepsilon_x \\ y_f - \varepsilon_y \\ \psi_f - \varepsilon_\psi \\ u_f - \varepsilon_u \end{bmatrix} \le \begin{bmatrix} x_f \\ y_f \\ \psi_f \\ u_f \end{bmatrix} \le \begin{bmatrix} x_f + \varepsilon_x \\ y_f + \varepsilon_y \\ \psi_f + \varepsilon_\psi \\ u_f + \varepsilon_u \end{bmatrix} \tag{6.41}$$

where $\cdot f$ denotes the optimal state and $\epsilon_.$ is the allowable tolerance for the corresponding state.

6.6.3 *Performance Analysis*

In order to verify the effectiveness of MOSFET RUL prediction result fusion for fault-tolerant control of wheeled robots. In this section, the MOSFET RUL prediction result data is used to simulate the faults during the wheeled mobile robot movement, which is used as a control constraint for wheeled mobile robot path planning. For this purpose, we developed a simulation program to design the multi-constraint control equations for wheeled mobile robots to accomplish global and local path planning (Table 6.3).

(1) The results show that different path planning methods have some scheduling planning capability. The method of fusing MOSFET RUL prediction results has a lower chain breakage rate compared to the path planning method alone. The possible reason is that fusing MOSFET RUL prediction results for fault-tolerant control can help the wheeled mobile robot control steering to reduce the excessive loss in the faulty region and ensure the task completion.
(2) From the action elapsed time, it can be seen that both CEEMD-LSTM-RRT and WPD-ARIMA-A* methods take more time than the original path planning algorithms. CEEMD-LSTM-RRT takes 0.783 s more time compared to RRT, and WPD-ARIMA-A* takes 3.319 s more time compared to A*. This is mainly due to the fusion of the RUL prediction result restricts the steering angle of the wheeled mobile robot, and in order to satisfy the traveling and obstacle avoidance requirements, the robot is unable to move along the optimal path leading to an increase in the elapsed time.
(3) WPD-ARIMA-A* has the longest action elapsed time but has a more stable traveling process compared to CEEMD-LSTM-RRT. The possible reason is that the MOSFET RUL prediction performance of WPD-ARIMA is better than that of CEEMD-LSTM, which makes the steering angle of the wheeled mobile robot relatively accurate in controlling the steering angle and does not produce complex fluctuations like CEEMD-LSTM-RRT (Figs. 6.9, 6.10, 6.11 and 6.12).

Table 6.3 Performance comparison of different path planning methods

Strategy	Operational time consumption (s)	Chain breakage rate (%)
RRT	33.855	0.0058
A*	34.172	0.0050
CEEMD-LSTM-RRT	34.638	0.0040
WPD-ARIMA-A*	37.491	0.0038

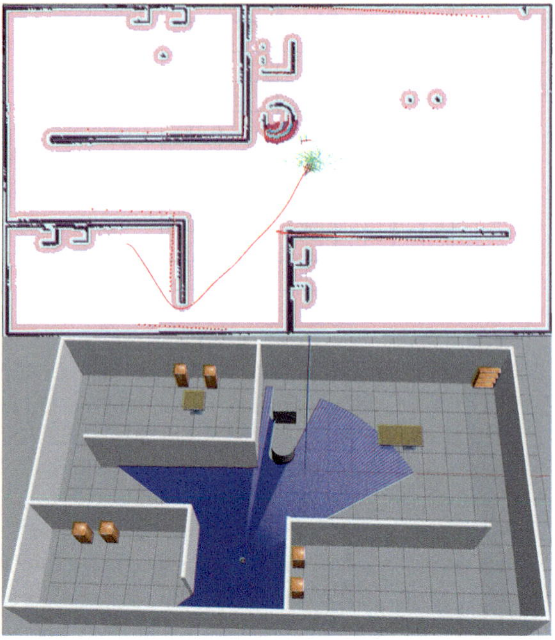

Fig. 6.9 RRT-based path planning for the wheeled mobile robot

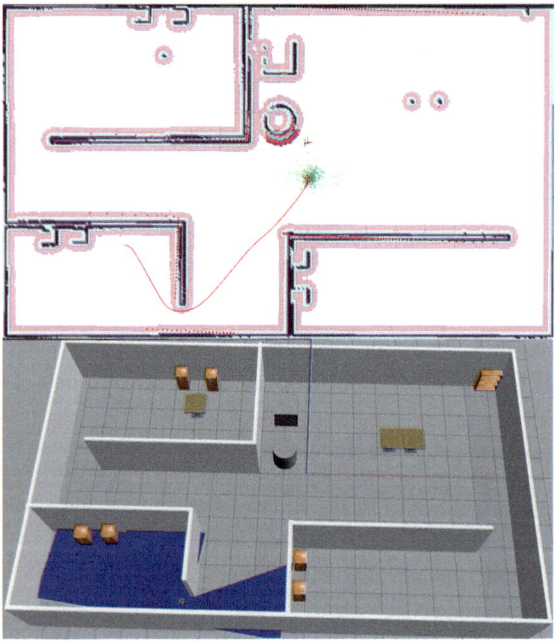

Fig. 6.10 CEEMD-LSTM-RRT-based path planning for the wheeled mobile robot

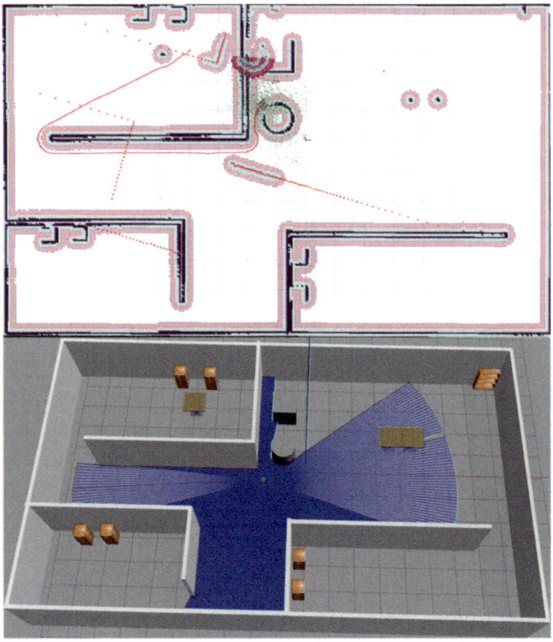

Fig. 6.11 A*-based path planning for the wheeled mobile robot

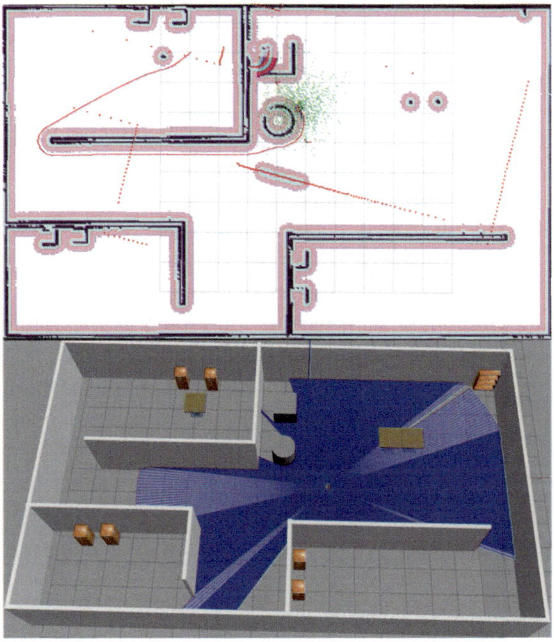

Fig. 6.12 WPD-ARIMA-A*-based path planning for the wheeled mobile robot

6.7 Conclusions

MOSFET remaining useful life prediction is critical for PHM in smart mechanical systems such as mobile robots, robotic arms, and drones, etc. MOSFETs, as critical components in these systems, affect their overall stable operation. Here are some potential applications of smart MOSFET RUL prediction techniques in PHM of smart mechanical systems:

(1) Predictive Maintenance and Optimization Strategies. During the working process of robots and other robots, the performance parameters of MOSFET are monitored in real time by sensors, and combined with the RUL prediction technology, preventive maintenance can be performed before the equipment fails, so as to reduce the unplanned downtime and improve the work efficiency. In addition, a more reasonable maintenance plan can be developed to avoid over-maintenance or under-maintenance, thus optimizing maintenance costs.
(2) Intelligent Decision Support and Remote Monitoring. Combining the RUL prediction technology with the control algorithm of the intelligent mechanical system can realize more intelligent decision support, such as automatic adjustment of working parameters and optimization of equipment performance. Remote monitoring of the real-time status of MOSFET, combined with RUL prediction technology can realize remote fault diagnosis and early warning to improve system reliability.
(3) Spare Parts Inventory Scheduling and Safety Prevention. Based on the MOSFET RUL prediction results, the demand and timing of spare parts can be predicted more accurately, thus reducing spare parts inventory and inventory costs. It can also detect and deal with potential MOSFET failures in a timely manner, avoiding safety accidents caused by equipment failures and safeguarding personnel and property.
(4) Digital Transformation and Smart Manufacturing. Applying RUL prediction technology to PHM of intelligent mechanical systems can promote digital transformation of industries and improve production efficiency and product quality. Combined with technologies such as the Internet of Matters, Big Data, and Artificial Intelligence, RUL prediction technology can promote the development of smart manufacturing and realize the intelligent, automated, and efficient operation of mobile robots and other equipment.

References

1. Celaya JR, Saxena A, Wysocki PF, Saha S, Goebel K (2010) Towards prognostics of power MOSFETs: accelerated aging and precursors of failure. In: Annual conference of the PHM Society
2. Chen GG, Zhang YZ, Huang J, Wang XZ, Kong XG (2024) A multi-source data fusion driven power field effect transistor health state assessment and remaining useful life prediction method. Meas Sci Technol 35(11). https://doi.org/10.1088/1361-6501/ad6f36

3. Choudhary K, Babu R, Christie L, Mahto M (2024) Real time conditioning monitoring of MOSFET using artificial neural network regression. Meas Sci Technol 35(8). https://doi.org/10.1088/1361-6501/ad4b50
4. Duan Z, Liu H (2019) An evolution-dependent multi-objective ensemble model of vanishing moment with adversarial auto-encoder for short-term wind speed forecasting in Xinjiang wind farm, China. Energy Convers Manag 198:111914
5. Dusmez S, Duran II, Akin B (2016) Remaining useful lifetime estimation for thermally stressed power MOSFETs based on ON-state resistance variation. IEEE Trans Ind Appl 52(3):2554–2563. https://doi.org/10.1109/TIA.2016.2518127
6. Fu SY, Tseng YC, Chiang K. N. Study on data effect of using RNN model to predict reliability life of wafer level packaging. In: 2020 15th international microsystems, packaging, assembly and circuits technology conference (IMPACT)
7. Kahraman CL, Roman D, Kirschbaum L, Flynn D, Swingler J (2024) Machine learning pipeline for power electronics state of health assessment and remaining useful life prediction. IEEE Access 12:136727–136746. https://doi.org/10.1109/ACCESS.2024.3460177
8. Li K, Xue YS, Cui S, Niu Q, Yang ZL, Luk P (2017) Advanced computational methods in energy, power, electric vehicles, and their integration
9. Li Z, Zheng Z, Outbib R (2018) A prognostic methodology for power MOSFETs under thermal stress using echo state network and particle filter. Microelectron Reliab 88–90:350–354. https://doi.org/10.1016/j.microrel.2018.07.137
10. Liang JY, Lu YX, Su MM (2024) Hga-lstm: LSTM architecture and hyperparameter search by hybrid GA for air pollution prediction. Genet Program Evolv Mach 25(2). https://doi.org/10.1007/s10710-024-09493-3
11. Pang JF, Luo W, Yao ZY, Chen J, Dong CY, Lin KR (2024) Water quality prediction in urban waterways based on wavelet packet denoising and LSTM. Water Resour Manage 38(7):2399–2420. https://doi.org/10.1007/s11269-024-03774-3
12. Ren HY, Yu YY, Liu JL, Zhou JJ, Du X (2023) Lifetime prediction of power MOSFET based on LSTM with successive variational mode decomposition and error compensation. IEICE Electron Express 20(16). https://doi.org/10.1587/elex.20.20230277
13. Rezaiy R, Shabri A (2024) An innovative hybrid W-EEMD-ARIMA model for drought forecasting using the standardized precipitation index. Nat Hazards 120(14):13513–13542. https://doi.org/10.1007/s11069-024-06758-z
14. Sang S, Li L (2024) A novel variant of lstm stock prediction method incorporating attention mechanism. MATHEMATICS 12(7). https://doi.org/10.3390/math12070945
15. Shen CK, Yu SY, Epureanu BI, Ersal Tulga (2023) An efficient global trajectory planner for highly dynamical nonholonomic autonomous vehicles on 3-D terrains. IEEE Trans Robot 40
16. Shin J, Shin JW, Kim W (2024) Noise-robust online measurement of the on-state resistance of the power semiconductor devices in PWM converters. IEEE Access 12:63669–63682. https://doi.org/10.1109/ACCESS.2024.3396575
17. Tian WQ, Chen NC (2024) Multiphysics coupling in IGBT modules: a review. J Electron Packag 146(4). https://doi.org/10.1115/1.4065941
18. Tsoku JT, Metsileng D, Botlhoko T (2024) A hybrid of Box-Jenkins ARIMA model and neural networks for forecasting South African crude oil prices. Int J Financ Stud 12(4). https://doi.org/10.3390/ijfs12040118
19. Wu LF, Yang J, Peng Z, Wang HM (2017) Remaining useful life prognostic of power metal oxide semiconductor field effect transistor based on improved particle filter algorithm. Adv Mech Eng 9(12). https://doi.org/10.1177/1687814017749324
20. Yan J, Li JH, Bai GX, Zong TC (2024) Identification of nonlinear system with time delay based on wavelet packet decomposition and Gaussian kernel GMDH network. Int J Syst Sci 55(8):1737–1753. https://doi.org/10.1080/00207721.2024.2317354
21. Yin S, Liu H, Duan Z (2021) Hourly PM2.5 concentration multi-step forecasting method based on extreme learning machine, boosting algorithm and error correction model. Digit Signal Process 118:103221

22. Zhang MY, Wang Q, Yu Y (2022) Health status identification and fault time prediction of MOSFET device based on LSTM-DHMM. Acta Electron Sin 50(03):643–651
23. Zhang Q, Gao JW, Dong HH, Mao YL (2018) WPD and DE/BBO-RBFNN for solution of rolling bearing fault diagnosis. Neurocomputing 312:27–33. https://doi.org/10.1016/j.neucom.2018.05.014
24. Zhao S, Chen SW, Yang F, Ugur E, Wang H (2020) A composite failure precursor for condition monitoring and remaining useful life prediction of discrete power devices. IEEE Trans Ind Inform PP (99):1–1
25. Zhou YM, Liu HZ, Mu SL, Chen ZQ, Wang B (2020) Short-circuit failure model of SiC MOSFET including the interface trapped charges. IEEE J Emerg Sel Topics Power Electron 8(1):90–98. https://doi.org/10.1109/JESTPE.2019.2939877

Chapter 7
Linear Networks and Temporal Convolution Based Prediction for Capacitor Remaining Useful Life

7.1 Introduction

As the new energy vehicles market rapidly expands, capacitors, as key electronic components, play an increasingly prominent role in vehicles [7]. Capacitors can store energy and provide instantaneous high current demands for vehicles, especially during startup and acceleration processes [1]. They can quickly release energy, acting as auxiliary batteries to provide additional power. Moreover, capacitors play a crucial role in filtering out high-frequency noise from power sources [16], ensuring the stability and cleanliness of the power supply, which is vital for protecting electronic devices from voltage fluctuations and electromagnetic interference [17]. Regarding power factor correction, capacitors enhance the efficiency of electrical energy utilization in on-board chargers and inverters, reducing the load on the power grid [2]. They also maintain voltage stability, especially during load changes, by quickly supplementing or releasing energy to keep the voltage constant [8]. Capacitors contribute to improving the electromagnetic compatibility of new energy vehicles by filtering out electromagnetic interference and protecting sensitive electronic devices inside the vehicle [18]. By smoothing power fluctuations and absorbing transient currents, capacitors help reduce battery stress, thereby extending their service life [27]. These functions of capacitors collectively enhance the overall performance of new energy vehicles [29], including power output, driving comfort, energy efficiency, and reliability [13].

The application of capacitors in new energy vehicles will become more extensive and in-depth [24], especially in the research and application of film capacitors and supercapacitors [4], providing critical support for the performance improvement and reliability assurance of new energy vehicles [15]. Film capacitors are primarily used for filtering and decoupling, reducing power supply noise and voltage fluctuations to ensure the stable operation of electronic systems. They are also suitable for high-frequency signal processing, optimizing the power conversion efficiency of inverters and motor drive systems. In contrast, supercapacitors excel in energy recovery, instant

H. Liu et al., *Prognostics and Health Management for Intelligent Electromechanical Systems*, https://doi.org/10.1007/978-981-96-7218-9_7

power demand, and start-stop systems. They can quickly capture recovery energy during braking to enhance the vehicle's range and fuel efficiency, while also rapidly releasing large amounts of energy during acceleration or climbing to supplement the main battery and improve vehicle performance. Additionally, supercapacitors can be combined with traditional batteries (such as lithium batteries) to balance the battery load, extend battery life, and improve overall system efficiency. Therefore, both types of capacitors have their unique strengths and work together to enhance the energy efficiency and performance of electric vehicles.

As more and more new energy vehicles adopt 800 V high-voltage architectures [22], high-voltage systems pose severe challenges to vehicle electronic components' service life and performance, especially capacitors [5]. Under high-voltage conditions, the voltage stress, operating temperature, and current fluctuations that capacitors withstand significantly increase [20], thus accelerating aging and failure [6]. In extreme conditions like high temperature and pressure, the dielectric material in capacitors can degrade, leading to issues such as decreased dielectric strength, increased leakage current, and a reduction in capacitance [28]. These factors can significantly impact the stability and safety of the entire vehicle [9].

Thus, examining how high voltage affects capacitor lifespan is essential for guaranteeing the reliable operation of new energy vehicles. A significant challenge in enhancing maintenance efficiency for these vehicles is the accurate prediction of capacitor remaining life using deep learning algorithms.

7.2 Data Acquisition

Accelerated aging tests are commonly employed to evaluate the degradation process for predicting capacitor lifespan. In this experiment, a high voltage from an external power source is applied to electrolytic capacitors, leading to continuous charging and discharging. NASA conducted similar tests on electrolytic capacitors and provided a corresponding degradation dataset [3]. The percentage of capacitance loss is illustrated in Fig. 7.1. This chapter uses capacitance loss percentage as a degradation indicator and develops an empirical model based on this data. For prediction purposes, Cap6 is selected, and due to the limited original data points, it is smoothed and sampled down to 194 points for the prediction dataset.

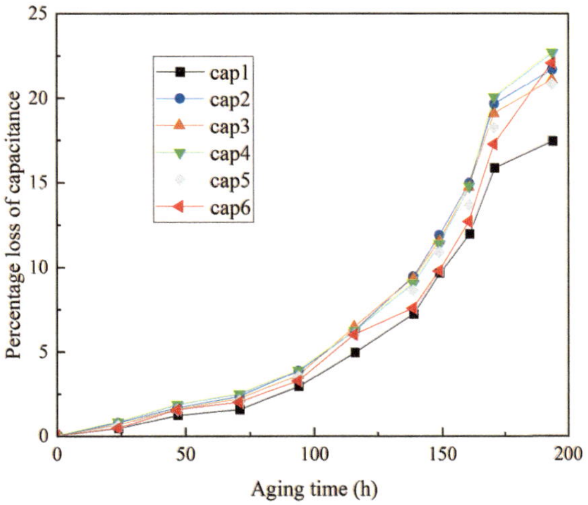

Fig. 7.1 Capacitance loss of six electrolytic capacitors (EOS dataset)

7.3 Predictive Model for Capacitor Remaining Useful Life Based on MSD-Mixer

7.3.1 The Theoretical Basis Linear Network

Most remaining life prediction models have been improved based on the Transformer framework. However, the self-attention mechanism of Transformers has excessively high computational complexity, and there is also the issue of position encoding not fully representing temporal information. Compared to this, linear network have the following advantages:

- Interpretability: The mathematical form is simple and intuitive, clearly showing the relationships between variables, which helps understand the dynamics behind time series data.
- Computational efficiency: The low computational cost makes them more efficient in handling large-scale time series data, especially in scenarios requiring real-time prediction or high-frequency updates.
- Stability: Linear Networks typically have a certain level of robustness against noise and outliers. They can handle outliers and noise relatively well, avoiding overfitting or producing unstable prediction results.

The linear network used in time series prediction research differs from MLP in combining multiple network modules. The basic framework is shown in Fig. 7.2.

In classical forecasting models, input time series $x(t)$ can be broken down into seasonal $s(t)$ and trend parts $f(t)$ and noise ϵ:

$$x(t) = s(t) + f(t) + \epsilon \tag{7.1}$$

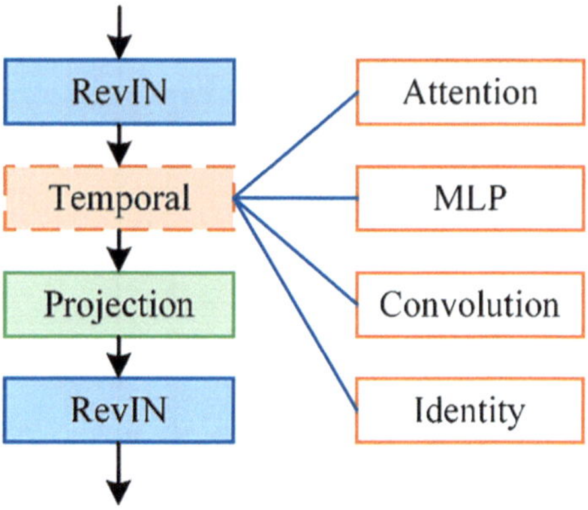

Fig. 7.2 Linear network

Theorem 1 *Given a seasonal time series where the period is $p \le n$, there is always an analytical solution to the Linear Network:*

$$[x_1, x_2, \ldots, x_n] \cdot \mathbf{W} + \mathbf{b} = \left[x_{n+1}, x_{n+2}, \ldots, x_{n+m}\right],$$
$$\mathbf{W}_{ij}^{(k)} = \begin{cases} 1, & \text{if } i = n - kp + (j \mod p) \\ 0, & \text{otherwise} \end{cases}, \quad 1 \le k \in \mathbb{Z} \le \lceil n/p \rceil, \quad b_i = 0. \tag{7.2}$$

Corollary 1 *If the given time series meets the condition $x(t) = ax(t-p) + c$ (where a and c are scaling and shifting factors, respectively), the Linear Network retains a closed-form solution:*

$$\mathbf{W}_{ij}^{(k)} = \begin{cases} a^k, & \text{if } i = n - kp + (j \mod p) \\ 0, & \text{otherwise} \end{cases}, \quad 1 \le k \in \mathbb{Z} \le \lceil n/p \rceil, b_i = \sum_{t=0}^{k-1} a^l \cdot c. \tag{7.3}$$

However, in practice, time series often follow classical models, and thus the trend component may affect the learning of Linear Network. So, handling the trend component is key to improving the learning efficiency of linear networks. The first step is to decompose the trend and seasonal features. Moving averages implemented by average pooling layers and sliding windows of appropriate size are used to extract trend information, and the trend component is subtracted from the original data to identify seasonal features in the periodic signal obtained. However, this approach may have the following issues:

- The sliding window size must exceed the maximum period of the seasonal component; otherwise, decoupling will be inadequate.
- An average pooling layer requires padding at both ends of the input time series, which can distort the beginning and end sequences.
- Even with a fully unraveled signal, there is a risk that the trend component may be underfitted.

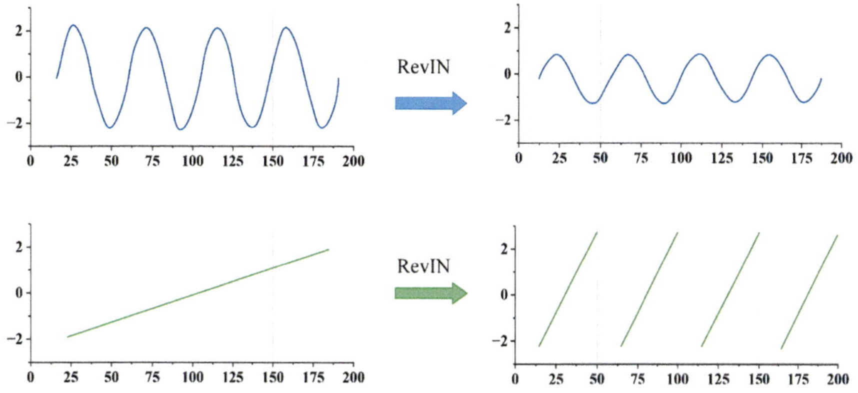

Fig. 7.3 Data decomposition of RevIN

After decoupling the trend component, reversible instance normalization transforms the trend into seasonality. Due to distribution shift issues, some statistical information of the time series, such as mean and variance, change over time. Reversible Instance Normalization (RevIN) standardizes the input historical time series first, inputs it into the prediction module, and finally denormalized the prediction result to transform the trend into seasonality in Fig. 7.3.

The capacitor data is normalized and denormalized by the RevIN module. The normalized sequence has a more consistent mean and variance. At the same time, the subsequent module can receive input for the consistent distribution of mean and variance. For each sequence of input data $x^{(i)} \in \mathbb{R}^{K \times T_x}$, the RevIN block calculation process is as follows:

$$\mathbb{E}_t\left[x_{kt}^{(i)}\right] = \frac{1}{T_x}\sum_{j=1}^{T_x} x_{kj}^{(i)} \tag{7.4}$$

$$\mathrm{Var}\left[x_{kt}^{(i)}\right] = \frac{1}{T_x}\sum_{j=1}^{T_x}\left(x_{kj}^{(i)} - \mathbb{E}_t\left[x_{kt}^{(i)}\right]\right)^2 \tag{7.5}$$

$$\hat{x}_{kt}^{(i)} = \gamma_k\left(\frac{x_{kt}^{(i)} - \mathbb{E}_t\left[x_{kt}^{(i)}\right]}{\sqrt{\mathrm{Var}\left[x_{kt}^{(i)}\right] + \varepsilon}}\right) + \beta_k \tag{7.6}$$

where, K is the number of variables, T_x is the length of the input sequence, T_y is the length of the input sequence, and $\gamma, \beta \in \mathbb{R}^K$ is the affine transformation parameter that can be learned. The denormalization method is as follows, using the same parameter values as in the normalization phase:

$$\hat{y}_{kt}^{(i)} = \sqrt{\mathrm{Var}\left[x_{kt}^{(i)}\right] + \varepsilon} \cdot \left(\frac{\tilde{y}_{kt}^{(i)} - \beta_k}{\gamma_k}\right) + \mathbb{E}_t\left[x_{kt}^{(i)}\right] \tag{7.7}$$

7.3.2 *The Theoretical Basis of MSD-Mixer*

MSD-Mixer [26] is a multi-scale decomposition model for time series analysis tasks, with the following characteristics:

- Multi-scale decomposition: MSD-Mixer employs a multi-layer MLP-Mixer module to decompose the time series into various components, each representing a different time scale. This approach enhances the model's ability to capture multi-scale time patterns within the data.
- Multi-scale timing patches: To effectively model multi-scale time patterns, MSD-Mixer adopt a timing patch method that divides the time series into non-overlapping patches, and uses MLP to capture changes within and between patches and channel correlations.
- Residual loss: MSD-Mixer proposed a new residual loss function, which constrained the mean and autocorrelation coefficients of decomposition residual to ensure the integrity of decomposition.

Figure 7.4 shows the overall architecture of MSD-Mixer. The MSD-Mixer is stacked from k layers by generating its low-dimension representation $\{E_1, \ldots, E_{\mathrm{k}}\}$ MICN, learn to decompose input X into k components $\{S_1, \ldots, S_{\mathrm{k}}\}$. In this case, $Z_0 = X$ is defined

$$Z_i = X - \sum_{j=1}^{i} s_j, \quad (i = 1, \ldots, k) \tag{7.8}$$

such that Z_i specifies the remaining part after the first i components have been decomposed from the input X, and we have

$$Z_i = Z_{i-1} - S_j. \tag{7.9}$$

As shown in Fig. 7.4, layer i of MSD-Mixer takes the remaining Z_{i-1} of the previous layer as input and learns to represent Z_{i-1} with a low-dimensional representation of E_i. The part represented is the Ith component S_i. More specifically, in each layer, Z_{i-1} is first processed in the Patching module and then input into the Patch Encoder module to generate the ith component $E_i = g_i(Z_i - 1)$. The Patch Decoder module then reconstructs the S_i from E_i, and the module restores it from the patch to the original dimension. After that, subtract S_i from Z_{i-1} to get Z_i. Z_i is fed as input into the next layer for further decomposition.

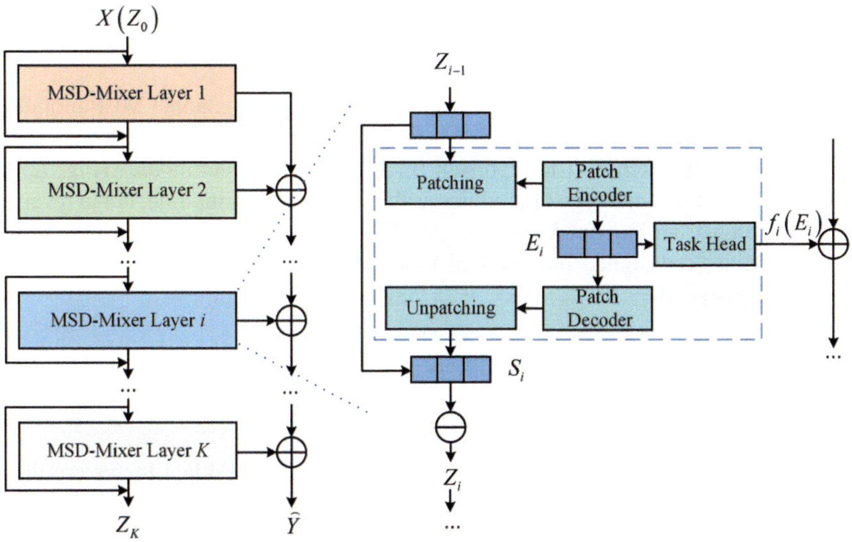

Fig. 7.4 Overall structure of MSD-Mixer

The Patch encoder and Patch decoder modules perform feature extraction based on MLPS of different dimensions. Figure 7.5a shows the structure of each MLP module. The following three types of MLP modules are used in the Patch encoder and Patch decoder modules:

- The Channel MLP module facilitates communication between different channels to capture inter-channel correlations.
- The Inter-Patch MLP module enables communication between various patches to capture global context.

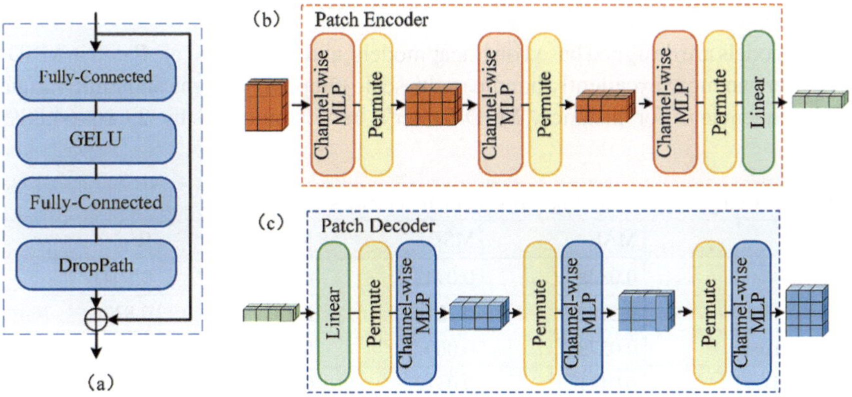

Fig. 7.5 Submodules of MSD-Mixer

- The MLP module within each Patch allows for communication between different time steps within the Patch, capturing changes at the sub-sequence level.

As shown in Fig. 7.5b, the Patch encoder is composed of channel MLP module, inter-patch MLP module, intra-patch MLP module and a linear layer to generate components from Patch Z_{i-1} to represent E_i. The Patch decoder module (Fig. 7.5c) has the same number and type of modules as the Patch encoder module, but in reverse order, to reconstruct S_i from E_i.

MSD-Mixer redesigned the residual Loss function. Firstly, the autocorrelation coefficient matrix $A = \{\, a_{i,j}\} \in \mathbb{R}^{C\times(L-1)}$ of Z_k is calculated:

$$a_{i,j} = \frac{\sum_{t=j+1}^{L}\left(z_{i,t} - \overline{z}_i\right)\left(z_{i,t-j} - \overline{z}_i\right)}{\sum_{t=1}^{L}\left(z_{i,t} - \overline{z}_i\right)^2} \tag{7.10}$$

$z_{i,j}$ represents the ith channel and the jth time step of Z_k. Residual losses are then defined by:

$$\mathcal{L}_r = \frac{\sum_{i,j} z_{i,j}^2}{C \times L} + \frac{\sum_{i,j}\left(\mathrm{ReLU}\left(\left|a_{i,j}\right| - \alpha/\sqrt{L}\right)\right)^2}{C \times (L-1)} \tag{7.11}$$

7.3.3 Experimental Result and Analysis

In this section, the Dlinear [23], Nlinear [23], MTS-Mixers [11], MSD-Mixer, and LightTS [25] models are used to predict the remaining service life of capacitors in high-voltage new energy vehicles. The model results are presented in Table 7.1 and illustrated in Fig. 7.6. By comparing the five models, the following comprehensive performance analysis of the MSD-Mixer can be derived:

- All models are designed based on linear models, and the multi-scale Patch in MSD-Mixer can more excellently handle multi-scale temporal patterns and multivariate dependencies. For instance, MSD-Mixer's MAE improved by 27.16–64.08%

Table 7.1 Comparison of linear network prediction metrics

Model	MAE	MSE	MAPE	R^2
Dlinear	0.0828	0.0948	0.3261	0.8941
Nlinear	0.1172	0.1214	0.4839	0.8328
MTS-Mixers	0.0578	0.0697	0.3210	0.9319
MSD-Mixer	0.0421	0.0512	0.2749	0.9476
LightTS	0.1019	0.1064	0.3991	0.8864

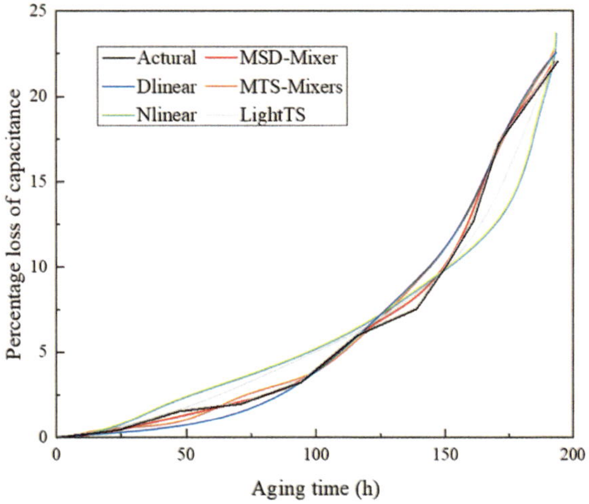

Fig. 7.6 Comparison of prediction effects of linear networks

compared to other models, indicating that MSD-Mixer's Patch communication mechanism is better at capturing the correlation between different features and extracting changes at the subsequence level.

- The way linear models handle different components can lead to significantly different prediction results. The linear layer decomposition strategy used in the DLinear model, which breaks down the time series into trend and residual components, is simple in structure but not particularly accurate. The Nlinear model shows advantages in handling distribution shifts and trend seasonality features, but the distribution shift in the original data is not severe. The time and channel factor decomposition strategy proposed by MTS-Mixers leverages the low-rank characteristics of time series data, offering high prediction accuracy. LightTS model is relatively simple, and the prediction accuracy needs further improvement. MSD-Mixer's approach to handling multi-scale Patches is more capable of extracting subtle changes in capacitor life data.

7.4 Predictive Model for Capacitor Remaining Useful Life Based on TimesNet

7.4.1 The Theoretical Basis of Temporal Convolutional Networks

Convolutional networks are primarily used for processing digital images and are essentially a special kind of convolution (filtering) operation conducted in the spatial domain. In contrast, temporal convolution operates on sequence data along the temporal dimension using causal convolution, meaning that the convolution operation is not conducted in the spatial domain but rather on sequence data over time.

Temporal convolutional network (TCN) employs several algorithmic optimizations:

- Dilated causal convolutions enable the convolutional kernel to cover a longer input range without increasing the number of parameters. This is achieved by inserting empty spaces, or dilation factors, within the kernel, effectively expanding the receptive field.
- Dilated convolutions (by inserting empty spaces/dilation factors in the convolutional kernel, allowing the kernel to cover a longer range of input without increasing the number of parameters, achieving an increased receptive field).
- Residual connections, which enable the construction of deeper networks and improve performance.

The structure of a TCN consists of multiple repeating modules, each containing two causal dilated convolutional layers and a residual connection. Each module has different dilation factors and the number of filters; an additional fully connected layer/softmax layer is added after the last module to output the prediction results.

TCN input and output data: The input data is a multidimensional array. For example, when processing biosignals, the first dimension represents the signal sequence length/time steps (depending on the sampling frequency), and the second dimension represents the number of features/signal channels (depending on how many channels and feature processing methods were collected). The output data is also a multidimensional array, with the first dimension still being the signal sequence length/time steps, and the second dimension representing the number of categories/ prediction targets (such as predicting the relationship between joint angles and time based on biosignals from multiple channels).

TCN adheres to two principles:

- Principle one: The input and output shapes of the TCN network must be identical to ensure that the model preserves information when processing temporal data.
- Principle two: The output at any given moment in the TCN network is convolutionally derived only from that moment and its preceding inputs, ensuring causality constraints when processing sequences.

TCN satisfies Principle two by employing causal convolution, a one-dimensional convolution operation that only considers past states. The specific steps of causal convolution are illustrated in Fig. 7.7, it has two characteristics:

- It only considers past information. The output at time y_t depends only on inputs $x_0, \ldots, x_t$, and not on any "future" inputs $x_{t+1}, \ldots, x_T$.
- The further back in history you go, the more hidden layers are required. In the diagram, if the output layer is expected to capture 5 time steps from the input layer, it requires 3 hidden layers. Different scenario requirements often demand taking into account more time steps.

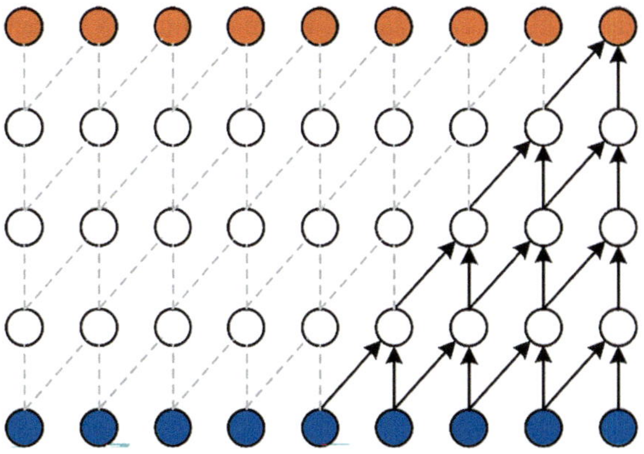

Fig. 7.7 Causal convolution

Due to the absence of recurrent connections, causal convolutions typically train faster than RNN, especially when applied to very long sequences. However, pure causal convolutions still suffer from the traditional convolutional neural network issue. To capture longer dependencies, a linear stacking of many layers is required. To address this issue, TCN use dilated convolutions to exponentially increase the receptive field without significantly increasing computational costs.

Unlike traditional convolutions, dilated convolutions, similar to dilated convolutions in CNNs, allow for spaced sampling during convolution, with the sampling rate controlled by the hyperparameter dilation, which refers to the subscript interval between elements within the kernel during the convolution operation.

The principle of dilated convolution is shown in Fig. 7.8, where blue circles represent the current convolution input time steps, orange circles represent the current convolution output time steps, white circles represent padding, and black arrows indicate the connection between input and output.

It can be seen that dilated convolution has three characteristics:

- The same characteristic as causal convolution, only considering past information.

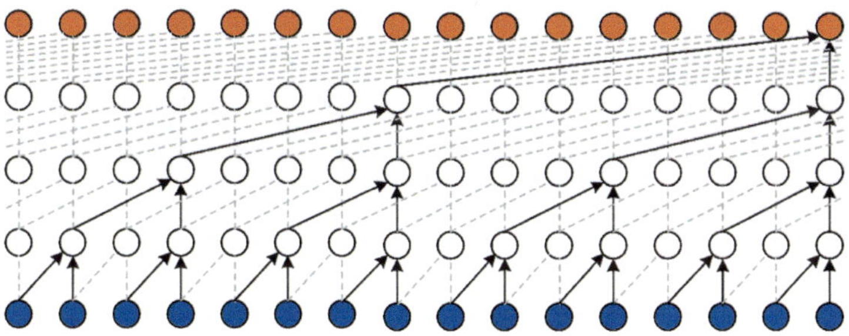

Fig. 7.8 Dilated convolution

- The same characteristic as dilated convolution, where each layer's extraction of information from the previous layer is skip-style.
- Each layer of dilated convolution requires padding (usually with zeros) to maintain the same dimension size. The size of the padding is: padding = (K − 1)*dilation.

Dilated convolutions enhance the scalability of causal convolutional networks, but the receptive field of TCN is still influenced by three parameters: network depth *n*, kernel size *K*, and dilation factor *d*. Even with the optimization of the number of network layers to some extent by dilated convolutions, a large number of layers are still required. The stability of such a deep network model is particularly important. For example, when predicting a situation that may depend on a history of size 212 and high-dimensional input sequences, a TCN network with at least 12 layers is needed, and each layer contains multiple convolutional kernels for feature extraction. Although increasing the network depth can extract more abstract and semantic features, it can also lead to gradient problems (vanishing or exploding gradients) and network degradation issues.

To address the aforementioned issues, TCN normalizes each weight vector and uses ReLU nonlinear activation functions and Dropout regularization. Additionally, residual connections are included in the design (residual connections are made every two dilated convolutions), as shown in Fig. 7.9.

The ReLU activation function and Dropout regularization not only help the network converge faster during training but also avoid the problem of vanishing gradients. Residual connections maintain the consistency of information flow and help the network learn identity mappings, thereby further improving network performance and generalization ability, and mitigating the issue of network degradation.

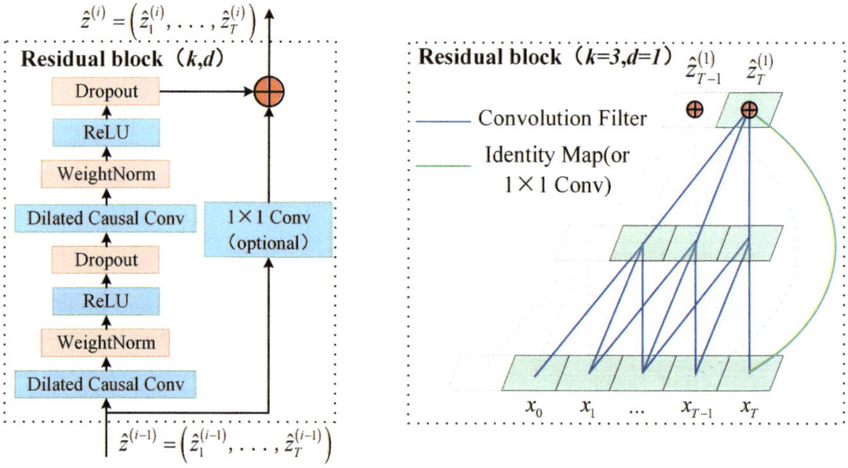

Fig. 7.9 TCN network model

The number of channels in the input and output of each TCN layer may be inconsistent, so an additional 1×1 convolution is required, and the results are added together time step by time step.

7.4.2 The Theoretical Basis of TimesNet

TimesNet first need to uncover the periodicity of the time series. For a one-dimensional time series $\mathbf{X}_{1D} \in \mathbb{R}^{T \times C}$ with a time length of T and a channel dimension of C, its periodicity can be calculated using the Fast Fourier Transform (FFT), that is:

$$\begin{aligned} \mathbf{A} &= \mathrm{Avg}(\mathrm{Amp}(\mathrm{FFT}(\mathbf{X}_{1D}))) \\ f_1, \ldots, f_k &= \underset{f_* \in \{1, \cdots, [\frac{T}{2}]\}}{\arg \mathrm{Topk}} (\mathbf{A}) \\ p_1, \ldots, p_k &= \left\lceil \frac{T}{f_1} \right\rceil, \ldots, \left\lceil \frac{T}{f_k} \right\rceil \end{aligned} \tag{7.12}$$

where $\mathbf{A} \in \mathbb{R}^T$ represents the intensity of each component in $\mathbf{X}_{1D}$, and the k frequencies $\{f_1, \cdots, f_k\}$ with the highest intensity correspond to the most significant k period lengths. The above process is briefly recorded as:

$$\mathbf{A}, \{f_1, \ldots, f_k\}, \{p_1, \ldots, p_k\} = \mathrm{Period}(\mathbf{X}_{\mathrm{1D}}) \tag{7.13}$$

Next, as shown in Fig. 7.10, the time series $\mathbf{X}_{1D}$ is folded. This process can be formalized as:

$$\mathbf{X}_{\mathrm{2D}}^i = \mathrm{Reshape}_{p_i, f_i}(\mathrm{Padding}(\mathbf{X}_{\mathrm{1D}})), i \in \{1, \cdots, k\} \tag{7.14}$$

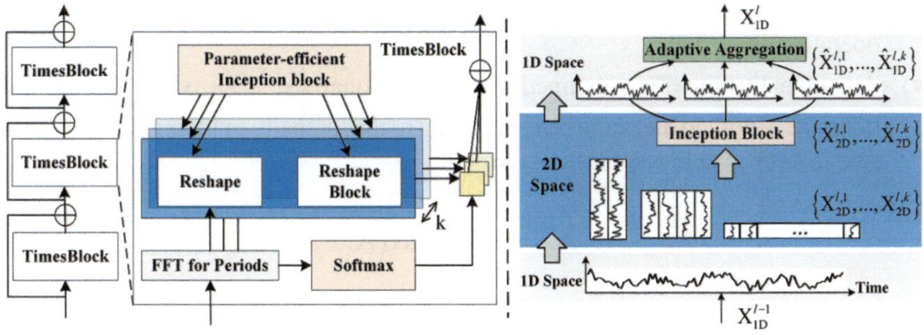

Fig. 7.10 TimesNet

In which, Padding(.) refers to adding zeros to the end of the sequence to ensure its length is divisible by p_i. This operation results in a set of two-dimensional tensors $\{\mathbf{X}_{2D}^1, \mathbf{X}_{2D}^2, \ldots, \mathbf{X}_{2D}^k\}$, $\mathbf{X}_{2\mathrm{D}}^i$ that correspond to the two-dimensional temporal changes governed by the period p_i.

TimesNet [21] is composed of stacked TimesBlocks. For the lth TimesBlock, its input is $\mathbf{X}_{1\mathrm{D}}^{l-1} \in \mathbb{R}^{T \times d_{\mathrm{model}}}$, and then two-dimensional temporal changes are extracted through 2D convolution:

$$\mathbf{X}_{1\mathrm{D}}^l = \mathrm{TimesBlock}\left(\mathbf{X}_{1\mathrm{D}}^{l-1}\right) + \mathbf{X}_{1\mathrm{D}}^{l-1} \tag{7.15}$$

Specifically, as shown in Fig. 7.10, The Timesblock contains four steps:

(1) One-dimensional transformation to two-dimensional: First, extract the period from the input one-dimensional temporal features $\mathbf{X}_{1D}^{l-1}$ and transform them into a two-dimensional tensor, which is the process described in the previous section:

$$\begin{aligned} &\mathbf{A}^{l-1}, \{f_1, \ldots, f_k\}, \{p_1, \ldots, p_k\} = \mathrm{Period}\left(\mathbf{X}_{1\mathrm{D}}^{l-1}\right) \\ &\mathbf{X}_{2\mathrm{D}}^{l,i} = \mathrm{Reshape}_{p_i, f_i}\left(\mathrm{Padding}\left(\mathbf{X}_{1\mathrm{D}}^{l-1}\right)\right), \quad i \in \{1, \ldots, k\} \end{aligned} \tag{7.16}$$

(2) Extraction of two-dimensional temporal change representation: For the two-dimensional tensor $\left\{\mathbf{X}_{2D}^{l,1}, \mathbf{X}_{2D}^{l,2}, \ldots, \mathbf{X}_{2D}^{l,k}\right\}$, which possesses two-dimensional locality, 2D convolution in this context, the classic Inception model is chosen, which includes:

$$\widehat{\mathbf{X}}_{2\mathrm{D}}^{l,i} = \mathrm{Inception}\left(\mathbf{X}_{2\mathrm{D}}^{l,i}\right) \tag{7.17}$$

(3) Transforming Two-Dimensional to One-Dimensional: For the extracted temporal features, they are converted back into one-dimensional space:

$$\widehat{\mathbf{X}}_{1\mathrm{D}}^{l,i} = \mathrm{Trunc}\left(\mathrm{Reshape}_{1,(p_i \times f_i)}\left(\widehat{\mathbf{X}}_{2\mathrm{D}}^{,ii}\right)\right), \quad i \in \{1, \ldots, k\} \tag{7.18}$$

where Trunc(.) denotes the removal of the zeros appended by the Padding(.) operation in step (1).

(4) Adaptive Fusion: The obtained one-dimensional $\left\{\widehat{\mathbf{x}}^{l,1}, \ldots, \widehat{\mathbf{x}}^{l,k}\right\}$ representations are weighted and summed according to the intensity of their corresponding frequencies to achieve the final output.

$$\begin{aligned} &\widehat{\mathbf{A}}_{f_1}^{l-1}, \ldots, \widehat{\mathbf{A}}_{f_k}^{l-1} = \mathrm{Softmax}\left(\mathbf{A}_{f_1}^{l-1}, \ldots, \mathbf{A}_{f_k}^{l-1}\right) \\ &\mathbf{X}_{1\mathrm{D}}^l = \sum_{i=1}^{k} \widehat{\mathbf{A}}_{f_i}^{l-1} \times \widehat{\mathbf{X}}_{1\mathrm{D}}^{l,i} \end{aligned} \tag{7.19}$$

7.4.3 Experimental Result and Analysis

In this section, the LSTNet [10], MICN [19], ModernTCN [14], SCINet [12], and TimesNet models are used to predict the remaining service life of capacitors in new energy electric vehicles, with the model results shown in Table 7.2 and Fig. 7.11. By comparing the five models, it can be concluded that TimesNet has the best overall performance. The experimental analysis is as follows:

All five models use temporal convolutional networks as the baseline model. TimesNet, utilizing Fourier transform, is more capable of capturing the multi-periodicity of time series data. For instance, the TimesNet model achieved an MAE of 0.324%, which is a 63.07–77.87% improvement compared to other models. SCINet is also a classic network that uses temporal convolution, but TimesNet's combination of frequency domain and time domain is more effective in extracting periodic changes.

The fusion of different types of networks does not necessarily achieve high prediction accuracy, while networks designed based on different periodic characteristics perform stronger. LSTNet combines convolutional and RNN networks, and ModernTCN combines convolutional and Transformer networks; although both

Table 7.2 Comparison of temporal convolution prediction metrics

Model	MAE	MSE	MAPE	R^2
LSTNet	0.1077	0.1431	0.4618	0.8462
MICN	0.1464	0.1566	0.5031	0.8251
ModernTCN	0.1348	0.1542	0.5849	0.7929
SCINet	0.0874	0.1034	0.3947	0.9237
TimesNet	0.0324	0.0671	0.3044	0.9540

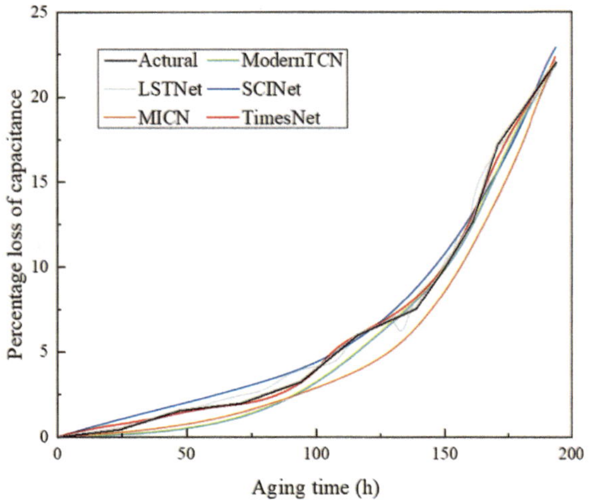

Fig. 7.11 Comparison of temporal convolution prediction results

leverage the advantages of two models, their overall prediction performance is not particularly impressive. MICN and SCINet both enhance the understanding of dependencies between different time periods through multi-scale methods. In particular, SCINet effectively models time series with complex temporal dynamics by obtaining intrinsic features at multiple resolutions through recursive downsampling-convolutions-interaction architecture. However, TimesNet's multi-period modeling allows the model to represent both intra-period and inter-period changes simultaneously, thereby better understanding and predicting time series data.

7.5 Model Performance Comparison

In this chapter, ten models include Dlinear, Nlinear, MTS-Mixers, MSD-Mixer, LightTS, LSTNet, MICN, ModernTCN, SCINet, and TimesNet were employed to test the models' predictive performance on the remaining service life of capacitors in new energy electric vehicles. This chapter explores the potential of Linear Networks and Temporal Convolutional models in the task of capacitor life prediction and analyzes the impact of different architectural designs. Based on the comparison of various predictive performance metrics in Fig. 7.12, the summary of the model analysis is as follows:

Linear Networks are more sensitive to capacitor life data compared to temporal convolutional models. The MAE of Linear Networks is 21.01% lower than that of temporal convolutional models. This may be because Linear Networks can effectively capture the basic linear relationships in the data, while temporal convolutional networks rely more on complex convolutional feature extraction processes, potentially overlooking some important linear trends. Capacitor life data often exhibit certain linear relationships, and Linear Networks can effectively capture this feature through simple linear regression, thus providing a lower MAE in small-scale datasets or in cases with less noise. In contrast, although temporal convolutional networks have advantages in dealing with complex time series data, their reliance on multi-layer convolution and activation function feature extraction processes may make the model less sensitive to subtle changes.

The fusion of multiple types of networks does not improve the prediction performance for capacitor life. The fusion of multiple types of networks is usually intended to enhance performance by combining the strengths of various models, but in practice, this strategy may not be effective for predicting capacitor life. The performance of LSTNet and ModernTCN networks is not good, possibly because there is redundancy in feature information between different network structures, leading to no significant gain in predicting capacitor life. Moreover, the merging of various model outputs may introduce additional noise, which can affect prediction performance. For tasks like capacitor life prediction, simpler and more effective models may be more suitable. Therefore, focusing on the optimization of a single structure and parameter tuning may be more effective than complex model fusion, which also emphasizes the importance of model selection and task feature adaptation.

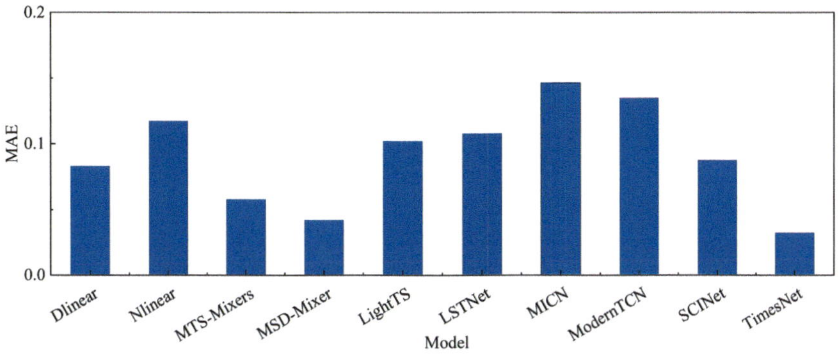

(a) MAE

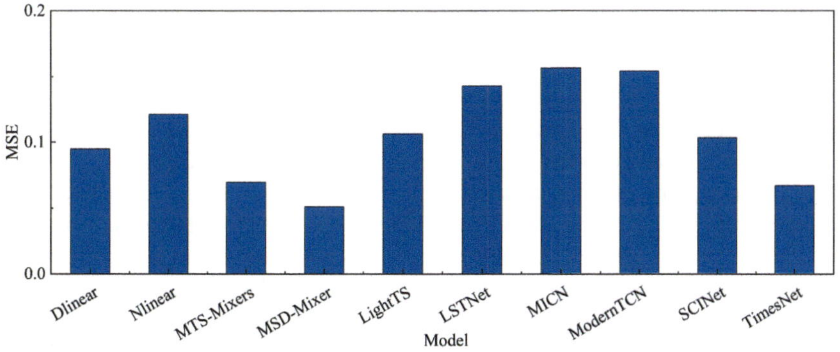

(b) MSE

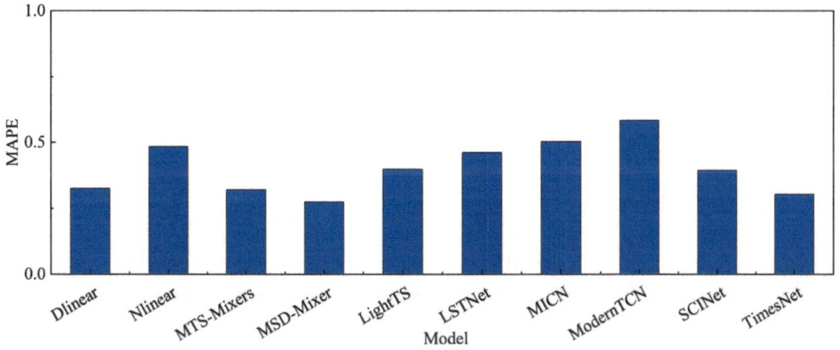

(c) MAPE

Fig. 7.12 Model performance comparison of all metrics

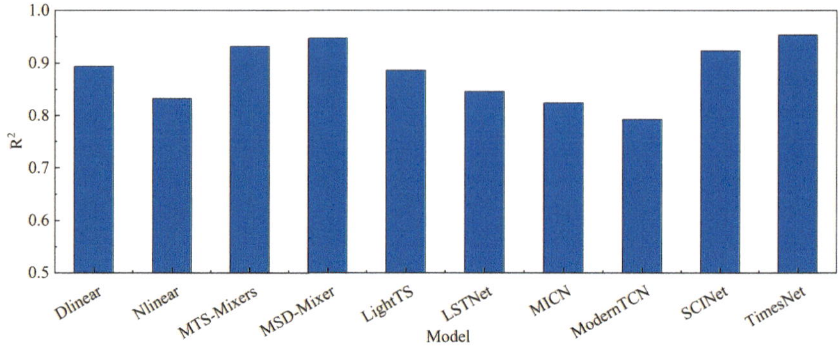

(d) R^2

Fig. 7.12 (continued)

Periodic analysis captures the intrinsic correlations of decay information better than data decoupling. The decay process of capacitors has certain cyclical factors, such as environmental temperature, operating frequency, and voltage changes, which affect the aging rate of capacitors. Periodic analysis helps researchers identify and analyze these cyclical changes and is very effective in capturing the patterns implicit in capacitor performance data. For example, TimesNet uses Fourier decomposition to obtain information from different time periods, and MSD-Mixer utilizes Patch partitioning to extract long-term dependencies from different sequence periods. In contrast, data decoupling techniques, although they can break down complex data into multiple parts, often fail to reveal the causal relationships and interdependencies between parts. For instance, Dlinear and Nlinear both divide the original data into trend and seasonal components. The seasonal component of capacitor life data is not significant.

7.6 Conclusions

With the integration of electrical components in new energy vehicles, capacitors, as a key component, have become a critical factor affecting the maintenance strategies of these vehicles. Current technologies for predicting the remaining service life of capacitors mainly focus on model parameter optimization, predictive algorithm architecture design, and the construction of capacitance physicochemical models. Despite certain achievements, the following areas still need to be strengthened:

Improve the prediction accuracy of universal model capacitors for different application scenarios. Since different application environments and working conditions significantly affect the performance of capacitors, their response characteristics in specific application scenarios need to be considered when predicting remaining life. For example, models and algorithms need to be developed for capacitors in extreme environments such as high temperature, high humidity, and vibration to obtain more accurate prediction results.

Study the impact of other electronic devices and key assemblies on the life of capacitors under the coupling effect of multi-physical fields. In the complex working environment of new energy vehicles, the life of capacitors is not only affected by their own conditions but also by the coupling effect of various factors in the overall vehicle system. For instance, temperature changes in the electric motor, electromagnetic interference from inverters, and current fluctuations in the on-board charging system can all impact capacitors. Therefore, in-depth coupled research on these factors is needed to determine their specific impact mechanisms and extent on the life of capacitors, thereby providing more comprehensive support for the reliability assessment of capacitors.

Enhance the generalization ability of algorithms by combining deep learning with the physicochemical characteristics of different structures of capacitors. Multi-level neural networks can be constructed to extract and learn features from data in combination with the physical properties of capacitors (such as ESR, capacitance, temperature coefficient, etc.). Meanwhile, considering the characteristic differences of different structures of capacitors (such as electrolytic capacitors, ceramic capacitors, etc.), targeted models can be designed to enhance the generalization ability of algorithms, making them applicable to the life prediction of capacitors of different types and brands.

References

1. Ayob A, Ansari S, Lipu MSH, Hussain A, Saad MHM, Ramees MKPM, Ahmad MW (2022) SOC, SOH and RUL estimation for supercapacitor management system: methods, implementation factors, limitations and future research improvements. Batteries 8(10):189. https://doi.org/10.3390/batteries8100189
2. Ayob A, Ansari S, Lipu MSH, Hussain A, Saad MHM, Ramees MKPM, Ahmad MW (2023) Advances in capacitor health monitoring techniques for power converters: a review. IEEE Access 11:133540–133576. https://doi.org/10.1109/ACCESS.2023.3336986
3. Celaya JR, Kulkarni CS, Biswas G, Goebel K (2012) Towards a model-based prognostics methodology for electrolytic capacitors: a case study based on electrical overstress accelerated aging. IJPHM 3(2). https://doi.org/10.36001/ijphm.2012.v3i2.1364
4. Chen JD, Miao JG, Yin Q (2024) A remaining useful life prediction method of aluminum electrolytic capacitor with adaptive degradation model selection. Microelectron Reliab 162:115509. https://doi.org/10.1016/j.microrel.2024.115509
5. Cui SH, Riaz S, Wang K (2021) Study on lifetime decline prediction of lithium-ion capacitors. Energies 16(22):7557. https://doi.org/10.3390/en16227557
6. Dave HB, Singh D, Bansal HO (2021) Multiple linear regression-based impact analysis of impedance network design on life expectancy of DC-link capacitor in q-ZSI fed motor drive. Eng Sci Technol Int J 24(1):171–182. https://doi.org/10.1016/j.jestch.2020.06.004
7. Gillani SW, Shahid K, Gulzar MM, Arif D (2024) Remaining useful life prediction of supercapacitors in electric vehicles using neural networks. Eng Appl Artif Intell 49(5):7327–7340. https://doi.org/10.1007/s13369-024-08766-4

8. Guo F, Lv HT, Wu XW, Yuan XH, Liu LL, Ye JL, Wang T, Fu LJ, Wu YP (2023) A machine learning method for prediction of remaining useful life of supercapacitors with multi-stage modification. J Energy Storage 73:109160. https://doi.org/10.1016/j.est.2023.109160
9. Kukurowski N, Pazera M, Witczak M (2024) Takagi-Sugeno observer design for remaining useful life estimation of Li-ion battery system under faults. Electronics 9(9):1537. https://doi.org/10.3390/electronics9091537
10. Lai GK, Chang WC, Yang YM, Liu HX (2018) Modeling long- and short-term temporal patterns with deep neural networks. In: SIGIR '18: the 41st international ACM SIGIR conference on research and development in information retrieval, 27 June 2018. ACM, Ann Arbor, MI, USA, pp 95–104. https://doi.org/10.1145/3209978.3210006
11. Li Z, Rao ZW, Pan LJ, Xu ZL (2023) MTS-Mixers: multivariate time series forecasting via factorized temporal and channel mixing. https://doi.org/10.48550/arXiv.2302.04501
12. Liu MH, Zeng AL, Chen MX, Xu ZJ, Lai QX, Ma LN, Xu Q (2022) Scinet: time series modeling and forecasting with sample convolution and interaction. Adv Neural Inf Process Syst 35:5816–5828
13. Lou GM, Lin WW, Huang GX, Xiang W (2023) A two-stage online remaining useful life prediction framework for supercapacitors based on the fusion of deep learning network and state estimation algorithm. Eng Appl Artif Intell 123:106399. https://doi.org/10.1016/j.engappai.2023.106399
14. Luo DH, Wang X (2024) ModernTCN: a modern pure convolution structure for general time series analysis. In: The twelfth international conference on learning representations
15. Ma N, Yin HX, Wang K, Gillani SW, Shahid K, Gulzar MM, Arif D, Lou G, Lin WW, Huang GX, Xiang W (2023) Prediction of the remaining useful life of supercapacitors at different temperatures based on improved long short-term memory. Energies 16(14):5240. https://doi.org/10.3390/en16145240
16. Qi GH, Ma N, Wang K (2024) Predicting the remaining useful life of supercapacitors under different operating conditions. Energies 17(11):2585. https://doi.org/10.3390/en17112585
17. Sawant V, Deshmukh R, Awati C (2023) Machine learning techniques for prediction of capacitance and remaining useful life of supercapacitors: a comprehensive review. J Energy Chem 77:438–451. https://doi.org/10.1016/j.jechem.2022.11.012
18. Wang CX, Xiong R, Tian JP, Lu JH, Zhang CM (2022) Rapid ultracapacitor life prediction with a convolutional neural network. Appl Energy 305:117819. https://doi.org/10.1016/j.apenergy.2021.117819
19. Wang HQ, Peng J, Huang FH, Wang JC, Chen JH, Xiao YF (2023) MICN: multi-scale local and global context modeling for long-term series forecasting. In: The eleventh international conference on learning representations
20. Wang XL, Jiang B, Ding SX, Lu NY, Li Y (2022) Extended relevance vector machine-based remaining useful life prediction for DC-Link capacitor in high-speed train. IEEE Trans Cybern 52(9):9746–9755. https://doi.org/10.1109/TCYB.2020.3035796
21. Wu HX, Hu TG, Liu Y, Zhou H, Wang JM, Long MS (2022) Timesnet: temporal 2d-variation modeling for general time series analysis. arXiv preprint arXiv:2210.02186
22. Yi ZX, Zhao K, Sun JR, Wang LC, Wang K, Ma YZ (2022) Prediction of the remaining useful life of supercapacitors. Math Probl Eng 2022:1–8. https://doi.org/10.1155/2022/7620382
23. Zeng AL, Chen MX, Zhang L, Xu Q (2023) Are transformers effective for time series forecasting? In: Proceedings of the AAAI conference on artificial intelligence, vol 9, pp 11121–11128
24. Zhang H, Yi ZX, Kang L, Zhang Y, Wang K (2024) A novel supercapacitor degradation prediction using a 1D convolutional neural network and improved informer model. Prot Control Mod Power Syst 9(4):51–68. https://doi.org/10.23919/PCMP.2023.000167
25. Zhang TP, Zhang YZ, Cao W, Bian J, Yi XH, Zheng S, Li J (2022) Less is more: fast multivariate time series forecasting with light sampling-oriented MLP structures. https://doi.org/10.48550/arXiv.2207.01186
26. Zhong SH, Song SZ, Zhuo WP, Li GY, Liu Y, Chan SHG (2024) A multi-scale decomposition MLP-Mixer for time series analysis. https://doi.org/10.48550/arXiv.2310.11959

27. Zhou YT, Huang YN, Pang JB, Wang K (2019) Remaining useful life prediction for supercapacitor based on long short-term memory neural network. J Power Sources 440:227149. https://doi.org/10.1016/j.jpowsour.2019.227149
28. Zhou YT, Wang YN, Wang K, Kang L, Peng F, Wang LC, Pang JB, Liu CL, Li DZ, Wang LC, Li LW, Wang K (2020) Hybrid genetic algorithm method for efficient and robust evaluation of remaining useful life of supercapacitors. Appl Energy 260:114169. https://doi.org/10.1016/j.apenergy.2019.114169
29. Zhou YT, Wang YN, Wang K, Kang L, Peng F, Wang LC, Pang JB, Liu CL, Li DZ, Wang LC, Li LW, Wang K (2022) Strong robustness and high accuracy in predicting remaining useful life of supercapacitors. APL Mater 10(6):061106. https://doi.org/10.1063/5.0092074

Chapter 8
Remaining Useful Life Prediction of Power Supply Based on Range-Extended New Energy Vehicles

8.1 Introduction

Range-extended new energy vehicles are a type of automobile that combines the advantages of electric vehicles and internal combustion engines [20]. These vehicles typically use an electric motor as the main propulsion system [5], equipped with a small internal combustion engine (usually gasoline or diesel). Under normal conditions, the vehicle primarily relies on battery power, driven by the electric motor, which offers high efficiency and low emissions. When the battery charge drops to a certain level, the internal combustion engine starts and acts as a generator to recharge the battery [21].

Range-extended electric vehicles increase the driving range of electric cars [24], addressing the limitations imposed by battery life in pure electric vehicles. Although range-extended electric vehicles still depend on internal combustion engines [16], their environmental impact is generally better than that of traditional fuel vehicles due to their high efficiency in electric mode and ability to reduce dependence on fossil fuels [7]. However, the unique power structure of range-extended vehicles, where the power battery is constantly subjected to charging and discharging, poses a significant challenge to battery life [1].

In range-extended electric vehicles, the parameter matching relationship between the battery and the range extender, as well as the overall vehicle energy management strategy, significantly impact the vehicle's fuel and electricity consumption, as well as the battery aging [12]. The requirement for pure electric driving range determines the size of the battery capacity match, and as the battery capacity increases [23], the proportion of the battery in the whole vehicle increases accordingly, which in turn increases the vehicle's energy consumption per 100 km and the overall lifecycle cost [15]. Additionally, different pure electric range battery parameter configurations may have varying impacts on battery aging, as battery aging is influenced by multiple factors [6], depth of discharge, and charge and discharge cycle frequency [11]. Therefore, it's essential to perform simulation analysis on the effects of battery

H. Liu et al., *Prognostics and Health Management for Intelligent Electromechanical Systems*, https://doi.org/10.1007/978-981-96-7218-9_8

aging and parameter matching across various driving ranges. The parameter matching relationship between the range extender and the battery also affects battery aging, as different sizes of range extenders can impact the battery's charge and discharge rates and frequency [17], thereby affecting battery aging. At the same time, to maintain bus voltage stability and prevent battery overcharge [18], an effective energy management strategy is needed to allocate energy reasonably among various power sources [24].

Various control strategies and vehicle conditions are selected for simulating the power battery [6]. Initially, power-following control is chosen to test the New European Driving Cycle (NEDC) and China Light Vehicle Test Cycle (CLTC) conditions of range-extended electric vehicles. The main methods are model-based, data-driven, and hybrid prediction methods [2]. Most of the studies describe the degradation behavior of the battery by fitting the capacity curve offline and then use various algorithms to predict the battery State of Health (SOH) curve, and finally calculate the battery RUL [14]. However, this method is not universal, and the prediction results are more random. In addition, data-driven methods do not have a fixed standard [8]. Short-term life prediction can only provide the battery SOH for a few future cycles, while long-term life prediction inevitably accumulates errors through the iterative process, leading to the final predicted battery SOH curve deviating from the actual battery SOH curve [4].

To address these issues, this chapter adopts a novel approach to predict the remaining life of the battery. First, the service environment of the power battery is determined, and the impact of its input and output physicochemical characteristics on the remaining life is analyzed. Then, a simulation environment for the power battery is built to truly simulate the service situation of the battery in range-extended vehicles and perform tests that meet the test standards. Finally, by integrating the established battery model with the multi-source data generated during the power battery's service process, an end-to-end learning model is developed to predict the battery's RUL.

8.2 Data Acquisition

To gather power battery data, a specific range-extended electric vehicle is used as a reference, utilizing the experimental data it provides for parameter matching and power system selection for the range-extended vehicle. A model of the range-extended vehicle was built, with the vehicle parameters as shown in Table 8.1.

The key indicators for assessing the power performance of electric vehicles include the 0–100 km/h acceleration time, maximum gradient during stable driving, and the maximum speed on good road conditions. The economic indicator is evaluated using the equivalent fuel consumption per 100 km. The power economy indicators necessary for the energy management study of the range-extended vehicle are presented in Table 8.2.

Table 8.1 Vehicle parameters

Vehicle parameter	Parameter value
Curb weight (kg)	1600
Gross weight (kg)	1975
Tire rolling radius (mm)	335 (215/55R18)
Calculated weight (kg)	1741.3
Track width (mm)	1585/1586
Wheelbase (mm)	2600
Center of gravity height (mm)	675.5
Distance from center of gravity to front wheel (mm)	1098

Table 8.2 Power economy indicators

Performance indicator	Value
0–100 km/h acceleration time (s)	≤ 10
Maximum speed (km/h)	≥ 140
Maximum gradient (%)	≥ 30
NEDC cycle fuel consumption (L/100 km)	≤ 9.1
CLTC cycle fuel consumption (L/100 km)	≤ 9.4

Through battery selection and parameter matching, the fundamental parameters of the power battery utilized in this chapter are displayed in Table 8.3.

Through engine selection and parameter matching, the basic parameters of the range extender used in this chapter are shown in Table 8.4.

This chapter employs a co-simulation platform using AVL_Cruise and Matlab/Simulink, and adopts the New European Driving Cycle (NEDC) and China Light Vehicle Test Cycle (CLTC) as the driving conditions for the simulation experiments. The vehicle model consists of multiple modules including a body module, wheel module, driver module, electric motor drive module, generator module, engine module, main reducer module, power battery module, and co-simulation module. As shown in Fig. 8.1.

Table 8.3 Basic parameters of the power battery

Parameter	Value
Battery type	Lithium-ion
Mass (kg)	0.225 + 0.015
Nominal voltage (V)	355.2
Cell voltage (V)	3.7
Total battery energy (kWh)	2.1
Cell capacity (Ah)	6
Battery operating range (%)	20–80

Table 8.4 Basic parameters of the range extender

Parameter	Value
Engine type	Four-cylinder four-stroke gasoline engine
Maximum engine power (kW)	75 (5800 rpm)
Maximum engine torque (Nm)	137 (4000 rpm)
Generator type	ISG integrated starter generator
Generator peak power (kW)	75
Generator peak torque (Nm)	130
Maximum speed (r/min)	12,000

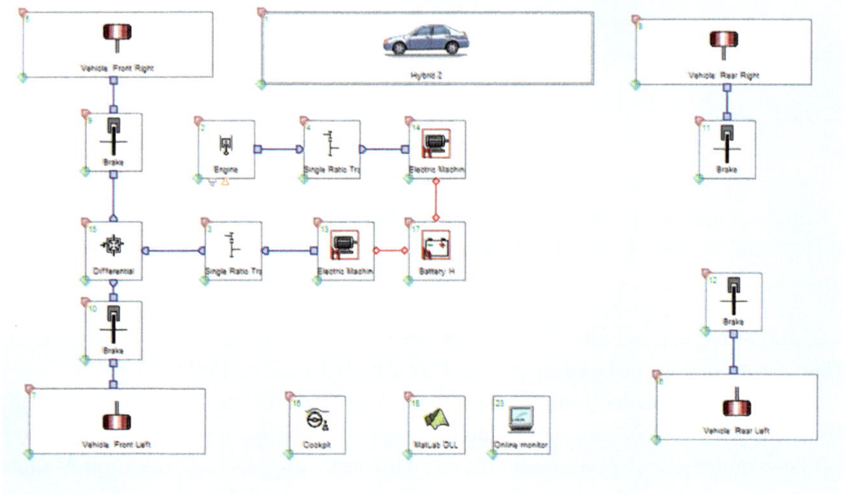

Fig. 8.1 Range-extended electric vehicle whole vehicle model

Input the selected parameters into Cruise to build models for the drive motor, power battery, and range extender. Enter the basic parameters of the power battery obtained from the matching and selection process into the power battery module. The relationship curve between SOC and voltage is input into the power battery module. The relationship curve between SOC and voltage of individual battery cells is shown in Fig. 8.2.

To verify the accuracy of power parameter matching and the construction of the whole vehicle model, it is necessary to establish simulation tasks to validate the power performance and economy. The energy management control strategy is a prerequisite for the simulation experiments of the vehicle's power and economy. This chapter uses Matlab/Simulink software to build the control strategy model. Due to the fast computation speed and strong decompilation capability of the Matlab_Dll module, co-simulation with the whole vehicle model is conducted through the Matlab_Dll module in AVL_Cruise.

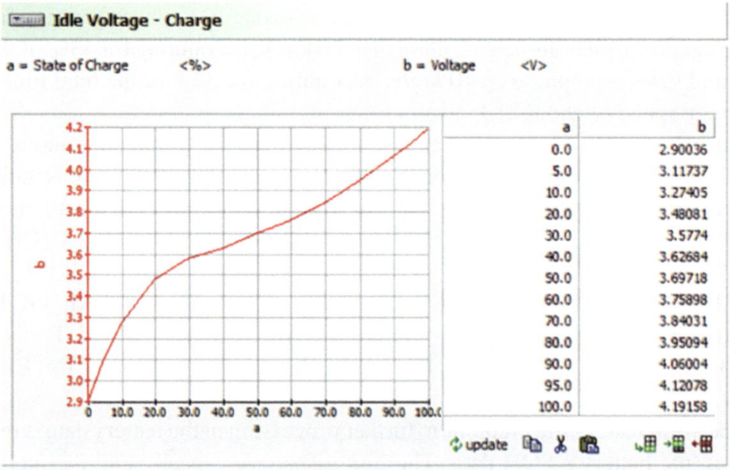

Fig. 8.2 Relationship between SOC and voltage of a battery

Next, the control strategy for the entire vehicle is established. The power-following control strategy is a widely used approach, where the key principle is to use the SOC threshold of the power battery as the switching condition for the driving mode of the range-extended vehicle. When switching to range-extended mode, the range extender adjusts to match the power required for the vehicle's travel. The specific implementation steps are as follows:

Based on the operating range of the power battery, SOC of 20% and 80% are used as the thresholds for whether the range extender should start. When the battery SOC drops to 20%, the vehicle switches to range-extended mode, and the range extender starts; when the range extender charges the power battery to SOC of 80%, the vehicle switches to pure electric mode, and the range extender is turned off, as shown below:

$$S(t) = \begin{cases} 0 & SOC \geq 80 \\ 1 & SOC \leq 20 \end{cases} \tag{8.1}$$

where $S(t)$ represents the current state of the engine, 0 indicates the engine is off, and 1 indicates the engine is on.

Finally, the test conditions for the entire vehicle are selected, including: NEDC and CLTC conditions. The NEDC condition includes two scenarios, simulating urban and suburban driving. Urban driving condition: a total of 780 s, simulating urban driving, including acceleration, maintaining speed, deceleration, and stopping, repeated four times. Suburban driving condition: starting from the 400th second, simulating suburban driving, with a significantly higher speed than the urban condition, tested for one cycle only. CLTC is a testing method that simulates the actual road driving environment in China. Its specific parameters include a total test time of 1800s (30 min) and a total driving distance of approximately 35 km. The test process is divided into three speed ranges: high-speed phase ($\geq$ 70 km/h, accounting for 10% of the total time), medium-speed phase (40–70 km/h, accounting for 55% of the total time), and low-speed phase (< 40 km/h, accounting for 35% of the total time), with an average speed of about 21 km/h.

To obtain sufficient data to describe battery life degradation, continuous NEDC and CLTC driving cycles are conducted, and power battery data is collected. There are a total of 15 types of working data for the power battery, including: TIME (s), VOLTAGE (V), CURRENT (A), CHARGE (Ah), STATE_OF_CHARGE (SOC %), MAXIMUM_POWER_OUTPUT (KW), MAXIMUM_POWER_INPUT (KW), ELECTRICAL_POWER (KW), POWER_LOSS (KW), TOTAL_OUTPUT_CHARGE (Ah), TOTAL_INPUT_CHARGE (Ah), TOTAL_OUTPUT_ENERGY (KJ), TOTAL_INPUT_ENERGY (KJ), SOURCE_VOLTAGE (V), INTERNAL_RESISTANT (Ohm).

To perform battery life prediction, further processing of the battery data is required to obtain the battery's SOH data. The loss of battery capacity and the increase in internal resistance can reflect the aging phenomenon of the battery. The expression for battery SOH is as follows:

$$SOH_{now} = \frac{C_{now}}{C_{New}} \times 100\% \tag{8.2}$$

$$SOH_{now} = \left(\frac{R_{EOL} - R_{now}}{R_{EOL} - R_{New}} \right) \times 100\% \tag{8.3}$$

where, *SOHnow* is the current health status of the battery, C_{now} is the maximum discharge capacity at the current cycle, *CNew* is the rated capacity of a new battery, *REOL*, *Rnow*, and *RNew* are the internal resistances at the end of life, the current moment, and a new battery, respectively.

In accordance with the IEEE1188-1996, IEEE1188-2005 standards and USABC guidelines, a battery cell is deemed to have reached the end of its usable life and requires replacement when its capacity degrades to 80% of the initial capacity or its internal resistance doubles from the initial level. The initial resistance of this power battery is 0.18 Ω, so a resistance value of 0.36 Ω is set as the marker for the end of the power battery's life. However, the estimation of SOH cannot be simply derived from the above formula, mainly because the SOH value is affected by the battery's working environment, material distribution, and structural design during the manufacturing process, and cannot be judged from a single indicator.

This chapter employs a commonly used method for estimating SOH based on open-circuit voltage (OCV) in actual processes: utilizing the resting voltage before charging data and the resting voltage after charging, OCV-SOC is used to calibrate SOC. The capacity charged, obtained by integrating the current over time, is divided by the change in SOC during charging ($\triangle$SOC), and the capacity of the entire battery pack can be derived. This is then compared with the rated capacity to calculate the current SOH. The formula for calculating the current battery capacity is as follows:

$$C_{now} = \frac{\int_{t_1}^{t_2} I(t)dt}{SOC(t_1) - SOC(t_2)} \tag{8.4}$$

where $SOC(t_i)$ represents the battery's charge level at time t_i, and $I(t)$ represents the instantaneous current of the battery at time t.

After calculating the battery SOH, the obtained range-extended electric vehicle battery data includes NEDC driving cycle data (with a data volume of 19,993 × 16) and CLTC driving cycle data (with a data volume of 22,663 × 16).

8.3 Predictive Model for Power Supply Remaining Useful Life Based on FEDformer

8.3.1 The Theoretical Basis of Transformer

Before the advent of the Transformer, RNN and LSTM were the mainstream methods for processing sequential data. However, these methods suffered from issues such as vanishing gradients and low computational efficiency when dealing with long sequences. To address these issues, the Transformer revolutionized the landscape of sequential data processing. The basic structure of the Transformer is the foundation of its powerful functionality, mainly consisting of positional encoding, encoders, and decoders, as shown in Fig. 8.3.

Positional Encoding: Since the Transformer architecture lacks recursive or convolutional structures, positional encoding is used to provide positional information. For the *pos*th position in the sequence, the value of the *i*th dimension of the absolute positional encoding vector is defined as follows:

$$\begin{aligned} PE_{(pos,2i)} &= \sin\left(pos/10000^{2i/d_{\text{model}}}\right) \\ PE_{(pos,2i+1)} &= \cos\left(pos/10000^{2i/d_{\text{model}}}\right) \end{aligned} \tag{8.5}$$

where d_{model} is the dimension of the input vector.

Encoder-Decoder Structure: The standard Transformer model employs an encoder-decoder structure. The encoder encodes the input sequence into contextual representations, and the decoder generates the output sequence based on these representations.

Self-Attention Layer: The Transformer calculates the relationships between each element in the input sequence through the self-attention mechanism. For an input sequence X, the Query, Key, and Value are obtained through linear transformations, then calculate the attention weights and perform a weighted sum, as shown in Fig. 8.4a:

$$Q = XW_Q, \quad K = XW_K, \quad V = XW_V \tag{8.6}$$

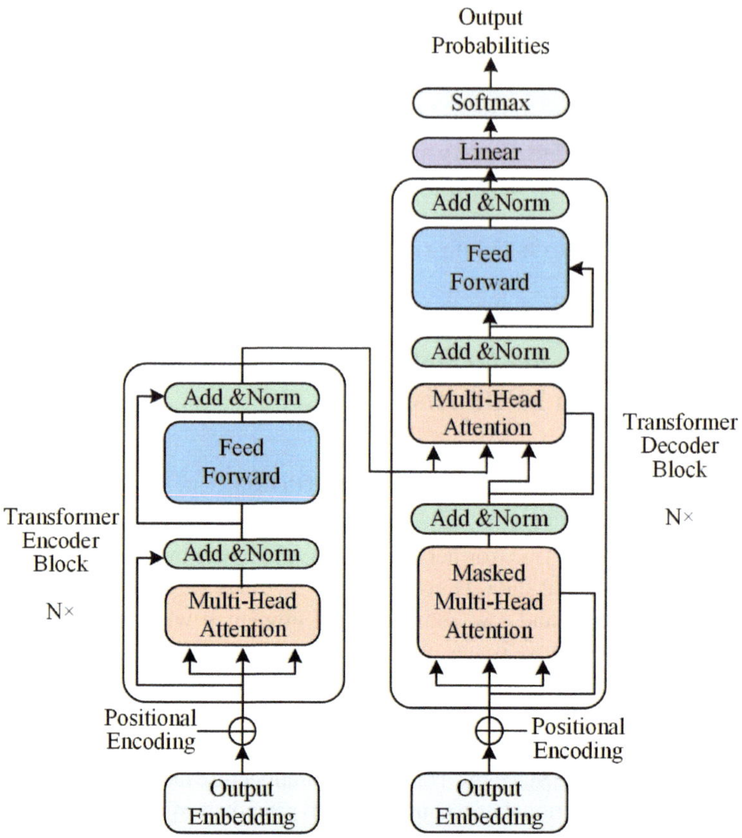

Fig. 8.3 Transformer

$$\text{Attention}(Q, K, V) = \text{softmax}\left(\frac{QK^T}{\sqrt{d_k}}\right)V \tag{8.7}$$

where W_Q, W_K, and W_V are learnable weight matrices, and d_k is the dimension of the keys, used for scaling to prevent excessively large values.

Each head can capture information from different subspaces of the input sequence in Fig. 8.4b:

$$\text{MultiHead}(Q, K, V) = \text{Concat}(head_1, \ldots, head_h)W^O \tag{8.8}$$

The computation for each head is:

$$\text{head}_i = \text{Attention}\left(QW_i^Q, KW_i^K, VW_i^V\right) \tag{8.9}$$

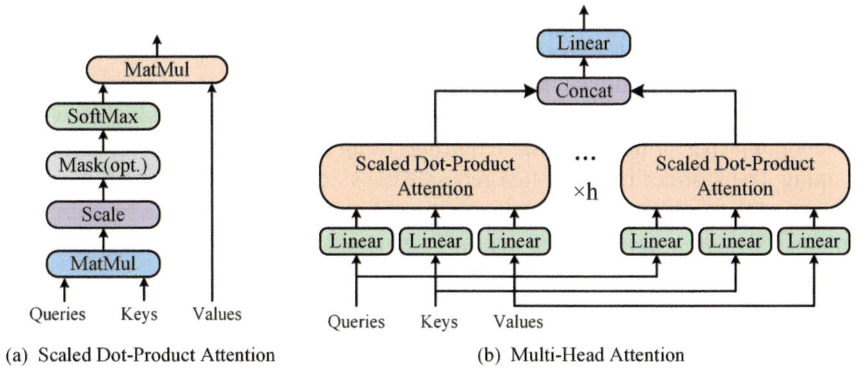

Fig. 8.4 Multi-head attention mechanism

Like W_Q, W_K and W_V matrices, W^O is a weight matrix trained together in the model training stage.

After the self-attention layer, the Transformer incorporates a position-wise feed-forward neural network. This component applies identical fully connected layers to each position separately, thereby enhancing the model's non-linear capabilities. Feed-forward network is shown below:

$$\text{FFN}(x) = \text{ReLU}(xW_1 + b_1)W_2 + b_2 \tag{8.10}$$

where W_1 and W_2 are randomly initialized weights, and b_1 and b_2 are biases for the linear transformations.

The Transformer is typically trained on large-scale datasets by optimizing the objective function (such as cross-entropy loss) to adjust model parameters. Common training strategies include using masking for decoder inputs to ensure that each position can only see the current and previous positions.

Transformers' efficient parallel processing capability stems from their multi-head self-attention mechanism. Each time step's computation is contingent upon the output of the previous step, hindering parallel computation, Transformers utilize self-attention to determine the representation of each position by computing its relevance with all other positions. This relevance calculation can be achieved through matrix multiplication, enabling parallel processing. The multi-head attention mechanism divides the input sequence into several subsequences, performs attention calculations on each, and then concatenates the results from different subsequences to obtain the representation of the entire input sequence. This approach allows for the simultaneous computation of relevance between multiple positions, effectively capturing long-range dependencies in sequences and significantly enhancing computational efficiency.

Transformers offer significant advantages in accuracy over early temporal processing models. This makes training and inference challenging in resource-constrained environments. Although capable of handling long-range dependencies, the computational time and memory consumption increase rapidly with longer input sequences, potentially leading to infeasibility.

Transformers exhibit inconsistent performance across datasets of varying scales. The self-attention mechanism lacks explicit sequential information, which may lead to suboptimal performance in tasks requiring strong sequence structure. Additionally, training Transformer models often requires considerable time, especially on large datasets. Due to the multitude of parameters, Transformers are prone to overfitting on small datasets, necessitating appropriate regularization and data augmentation strategies.

In summary, while the advantages of Transformers in sequence data processing make them the foundation of many state-of-the-art technologies, their limitations must be weighed in specific scenarios.

8.3.2 The Theoretical Basis of FEDformer

FEDformer [27] is a model designed for time series forecasting. Its structure integrates a frequency enhancement mechanism with the Transformer architecture, primarily by decomposing sequence information to better capture features at different frequencies, as shown in Fig. 8.5.

The main backbone of FEDformer, employs an encoder-decoder architecture and includes four sub-modules: the Frequency Enhanced Block, the Frequency Enhanced Attention module, the MOE Decomp module, and the Feed Forward module.

Within the encoder, the input passes through two MOE Decomp layers, each of which decomposes the signal into seasonal and trend components. The trend component is discarded, while the seasonal component is handed over to the subsequent layers for learning and eventually passed to the decoder. Given a sequence of battery data $\mathcal{X}_{\text{en}}^{l} \in \mathbb{R}^{I \times D}$, where I represents the length of the sequence and D represents the dimension of the sequence data, the computation in the encoder is as follows:

$$\mathcal{S}_{\text{en}}^{l,1}, - = \text{MOEDecomp}\left(\text{FEB}\left(\mathcal{X}_{\text{en}}^{l-1}\right) + \mathcal{X}_{\text{en}}^{l-1}\right)$$

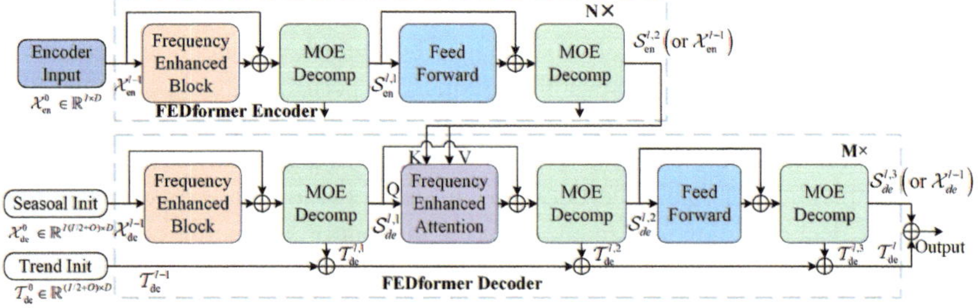

Fig. 8.5 FEDformer

$$\mathcal{S}_{\text{en}}^{l,2}, - = \text{MOEDecomp}\left(\text{FeedForward}\left(\mathcal{S}_{\text{en}}^{l,1}\right) + \mathcal{S}_{\text{en}}^{l,1}\right)$$
$$\mathcal{X}_{\text{en}}^{l} = \mathcal{S}_{\text{en}}^{l,2} \tag{8.11}$$

where $\mathcal{S}_{\text{en}}^{l,1}$, $i \in \{1, 2\}$ represents the seasonal component after the ith decomposition block in layer l.

In the decoder, the input from the encoder also passes through three MOE Decomp layers, which decompose the signal into seasonal and trend components. The seasonal component is then passed to subsequent layers for further learning. During this process, the Frequency Enhanced Attention layer is used to learn the frequency-domain correlations between the seasonal terms of the encoder and decoder. The trend component is accumulated and eventually added back to the seasonal component to reconstruct the original sequence. The computation in the decoder is as follows, $\mathcal{X}_{\text{de}}^{l}, \mathcal{T}_{\text{de}}^{l} = \text{Decoder}(\mathcal{X}_{\text{de}}^{l-1}, \mathcal{T}_{\text{de}}^{l-1})$ where $l \in \{1, \ldots, \text{N}\}$:

$$\mathcal{S}_{\text{de}}^{l,1}, \mathcal{T}_{\text{de}}^{l,1} = \text{MOEDecomp}\left(\text{FEB}\left(\mathcal{X}_{\text{de}}^{l-1}\right) + \mathcal{X}_{\text{de}}^{l-1}\right)$$
$$\mathcal{S}_{\text{de}}^{l,2}, \mathcal{T}_{\text{de}}^{l,2} = \text{MOEDecomp}\left(\text{FEA}\left(\mathcal{S}_{\text{de}}^{l,1}, \mathcal{X}_{\text{cn}}^{N}\right) + \mathcal{S}_{\text{de}}^{l,1}\right)$$
$$\mathcal{S}_{\text{de}}^{l,3}, \mathcal{T}_{\text{de}}^{l,3} = \text{MOEDecomp}\left(\text{FeedForward}\left(\mathcal{S}_{\text{de}}^{l,2}\right) + \mathcal{S}_{\text{de}}^{l,2}\right)$$
$$\mathcal{X}_{\text{de}}^{l} = \mathcal{S}_{\text{de}}^{l,3}$$
$$\mathcal{T}_{\text{de}}^{l} = \mathcal{T}_{\text{de}}^{l-1} + \mathcal{W}_{l,1} \cdot \mathcal{T}_{\text{de}}^{l,1} + \mathcal{W}_{l,2} \cdot \mathcal{T}_{\text{de}}^{l,2} + \mathcal{W}_{l,3} \cdot \mathcal{T}_{\text{de}}^{l,3} \tag{8.12}$$

where $\mathcal{S}_{\text{de}}^{l,i}, \mathcal{T}_{\text{de}}^{l,i}$, $i \in \{1, 2, 3\}$ represent the seasonal and trend components, respectively, after the ith decomposition block at the lth layer. $\mathcal{W}_{l,i}$, $i \in \{1, 2, 3\}$ represents the projector for the ith extracted trend $\mathcal{T}_{\text{de}}^{l,i}$.

The Frequency Enhanced Block and Frequency Enhanced Attention are the main components of the FEDformer. As depicted in Fig. 8.6a, the primary computational steps are as follows: initially, the time-domain input sequence $x \in \mathbb{R}^{\text{N}\times D}$ is linearly projected through an MLP, then the query q is transformed from the time domain to the frequency domain $Q \in \mathbb{C}^{N\times D}$, and finally, Q is subjected to random sampling within the frequency domain $\tilde{Q} \in \mathbb{C}^{M\times D}$.

$$\widetilde{\mathbf{Q}} = \text{Select}(\mathbf{Q}) = \text{Select}(\mathcal{F}(\mathbf{q})) \tag{8.13}$$

Here, M denotes the number of random samplings performed.

Subsequently, parameters $R \in \mathbb{C}^{M\times D\times D}$ are initialized, and through parameter learning, $\tilde{Y} \in \mathbb{C}^{M\times D}$ is generated. To restore the signal to its original length, it is necessary to pad the frequency points that were not sampled in step 2 with zeros to obtain $Y \in \mathbb{C}^{N\times D}$. Ultimately, the sequence is transformed back into the time domain to generate the reconstructed signal $y \in \mathbb{R}^{L_{de}\times D}$.

Define $q \in \mathbb{R}^{L\times D}$, $k \in \mathbb{R}^{L\times D}$, and $v \in \mathbb{R}^{L\times D}$ as elements taken from the decoder, and q is taken from the decoder's $q = x_{en} \cdot w_q$, where $w_q \in \mathbb{R}^{D\times D}$. The keys k

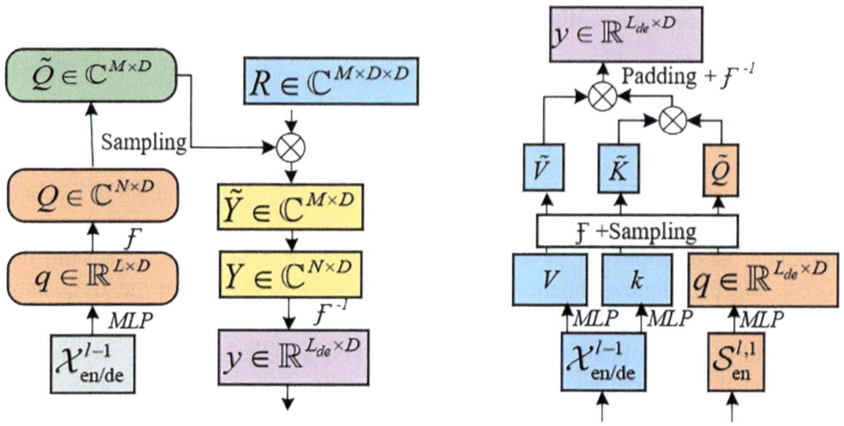

(a) Frequency Enhanced Block (b) Frequency Enhanced Attention

Fig. 8.6 Submodule of FEDformer

and values v are taken from the decoder's $k = x_{de} \cdot w_k$ and $k = x_{de} \cdot w_k$, respectively, where $w_k, w_v \in \mathbb{R}^{D \times D}$. The standard attention mechanism is then transformed into Frequency Enhanced Attention, as shown in Fig. 8.6b. Below is a comparison between the two attention mechanisms:

$$\text{Attention}(Q, K, V) = \text{softmax}\left(\frac{QK^T}{\sqrt{d_k}}\right)V \tag{8.14}$$

$$\begin{aligned} \widetilde{Q} &= \text{Select}(\mathcal{F}(q)) \\ \widetilde{K} &= \text{Select}(\mathcal{F}(k)) \\ \widetilde{V} &= \text{Select}(\mathcal{F}(v)) \\ \text{FEA}(q, k, v) &= \mathcal{F}^{-1}\left(\text{Padding}\left(\text{softmax}\left(\widetilde{Q} \cdot \widetilde{K}^{\top}\right) \cdot \widetilde{V}\right)\right) \end{aligned} \tag{8.15}$$

The model addresses the issue of inaccurate predictions due to the significant distribution discrepancy between the model's inputs and true values. The FEDformer mitigates this distribution difference between inputs and outputs by employing seasonal-trend decomposition, and it incorporates a frequency-domain attention mechanism into its model structure to enhance robustness against noise.

8.3.3 Experimental Result and Analysis

In this section, Autoformer [22], FEDformer, Informer [26], N-Transformer [10], and Transformer models are used to predict the remaining useful life of batteries under two different test conditions: CLTC and NEDC. The results of the models are illustrated in Table 8.5 and Fig. 8.7. A comparison among the five models reveals that the FEDformer demonstrates the most superior overall performance. The detailed experimental analysis is provided as follows.

All models are designed based on the Transformer as a benchmark model. The frequency domain learning mechanism of the FEDformer is more sensitive to long-term battery life data. For instance, the Mean Absolute Error (MAE) of the FEDformer in CLTC and NEDC conditions improved by 62.03–48.66% compared to other models. This indicates that the FEDformer is more capable of capturing the intrinsic characteristics of how battery life signals change over time.

All five models modify the multi-head self-attention mechanism and perform simple decomposition of sequences. The Autoformer employs a novel decomposition framework with an automatic relevance mechanism, but the model is relatively simple and not very sensitive to long sequences. The Informer model focuses more on reducing computational complexity and is not particularly outstanding in terms of accuracy. In comparison, the FEDformer, combined with frequency domain methods, shows greater potential in terms of life prediction capabilities.

Table 8.5 Comparison of predictive indicators based on FEDformer

Model	Dataset	MAE	MSE	MAPE	R^2
Autoformer	CLTC	0.3677	0.3861	1.4293	0.7644
	NEDC	0.2783	0.3124	1.6433	0.7863
FEDformer	CLTC	0.1266	0.1235	0.8946	0.8947
	NEDC	0.1541	0.0946	0.8732	0.9496
Informer	CLTC	0.3334	0.3278	1.2483	0.7946
	NEDC	0.3328	0.2946	1.2481	0.7569
N-Transformer	CLTC	0.2984	0.2743	1.3122	0.7356
	NEDC	0.3467	0.3159	1.2423	0.7469
Transformer	CLTC	0.2466	0.2496	0.9356	0.8634
	NEDC	0.1894	0.2236	0.9984	0.8443

Fig. 8.7 Comparison of prediction results based on FEDformer

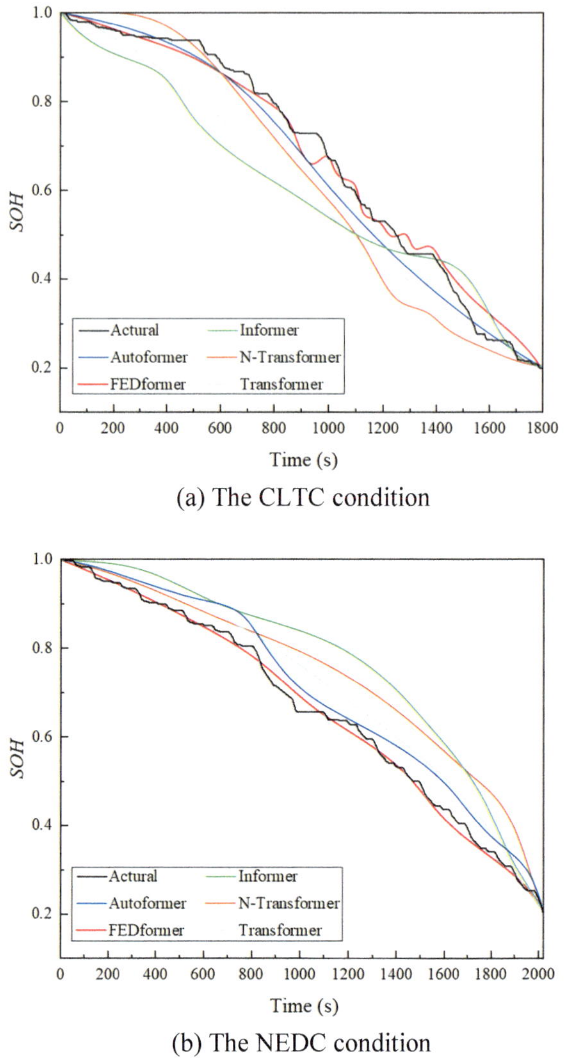

8.4 Predictive Model for Power Supply Remaining Useful Life Based on Preformer

8.4.1 The Theoretical Basis of Multi-scale Time–Frequency Analysis of Power Batteries

The theory of multi-scale extraction for power battery time series data is primarily used to analyze the dynamic data generated during the charging and discharging processes of batteries, including voltage, current, temperature, capacity, and internal impedance, among other indicators. Batteries exhibit complex characteristics in terms of multi-scale, multi-level, multi-process, multi-step, and multi-field coupling across spatial and temporal dimensions.

First, let's analyze the characteristics of power battery time series data:

Nonlinearity and non-stationarity. The performance indicators of batteries exhibit complex nonlinear characteristics under different operating states (charging, discharging, etc.).

Periodicity and transient response. The voltage and current signals of batteries are influenced by periodic charging and discharging processes and transient responses.

Environmental impact: Factors such as temperature and humidity significantly affect battery performance, leading to data complexity.

Traditional multi-scale studies are based on sequence decomposition, mainly involving wavelet transforms and Fourier transforms. Wavelet transforms can decompose signals into different frequency components. This decomposition is achieved through a set of basis functions known as wavelets, which, unlike traditional Fourier transforms, offer good time–frequency localization properties. Scholars have developed variants of wavelet transforms tailored to different characteristics of time series data, including CWT, DWT, Energy Concentration Wavelet Transform (ECWT), and WPT. Each variant has its own characteristics and is suitable for different application scenarios. For example, CWT and DWT are widely used in signal processing, ECWT is suitable for transient signal analysis requiring high time–frequency resolution, and WPT excels in occasions demanding more precise signal feature extraction.

Fourier transforms include Discrete Fourier Transform (DFT), FFT, Continuous Fourier Transform (CFT), and STFT. DFT is the Fourier transform for discrete signals and is particularly suitable for computer processing and analysis. FFT is an efficient algorithm for DFT, significantly reducing computational complexity. It is commonly used in practical applications to handle large-scale time series data. Wavelet transform is another time–frequency analysis method that, although not a Fourier transform, complements Fourier analysis. It allows for the analysis of signals at different time and frequency scales and is highly suitable for non-stationary time series data. The various forms and methods of Fourier transforms are crucial in numerous application scenarios. The selection of the appropriate transformation method typically hinges on the characteristics of the data and the specific analysis needs. In recent years, as artificial intelligence has gained momentum.

In recent years, integrating multi-scale analysis into time series forecasting models has been shown to effectively enhance the accuracy and robustness of predictions. On one hand, different features have varying periodic patterns that require characterization at different scales. On the other hand, smaller scales offer finer detail but come with increased computational complexity, while larger scales are more coarse and reduce computational load, presenting a trade-off to be balanced. Identifying the optimal modeling scale for different time series features and establishing interactive relationships among multi-features and multi-scales directly determines the model's performance. This work includes: data embedding, patch strategies, and attention mechanism design methods (within patches, between patches, and among multi-features).

Data embedding mainly involves: feature selection and dimensionality reduction. In data preprocessing, methods such as PCA, t-SNE, or other dimensionality reduction techniques are used to select important features, thereby reducing redundant information; temporal feature embedding. Time encoding (e.g., using sine and cosine functions for positional embedding of time) allows the model to capture periodic time patterns; segmented embedding. For periodic features, they can be processed in segments, with each cycle's features embedded separately to obtain more detailed cyclic information.

Patch strategies include local time segment extraction and dynamic patch size adjustment. Local time segment extraction: Long time series data are divided into multiple smaller segments (patches), with each patch modeled independently. This method can capture fine-grained information at different time scales while reducing the computational complexity for a single model. Larger patches can identify long-period patterns, while smaller patches are better at recognizing high-frequency patterns. For example, time series data can be adaptively divided into optimal, different-scale patches using Fourier transform, and attention mechanisms are designed within and between patches to capture global and local dependencies. Dynamic patch size adjustment: Patch sizes are adjusted dynamically based on data characteristics. Smaller patches are used for parts with rapid changes, and larger patches for parts with slow changes, allowing for detailed capture of details while effectively reducing computational load.

Attention mechanism design includes intra-patch attention mechanisms, inter-patch attention mechanisms, and multi-feature attention. Intra-patch attention mechanisms: Self-attention mechanisms are used within each patch, enabling the model to automatically determine which time step in that feature channel is most important for the current prediction. This allows the model to focus on locally important information. Inter-patch attention mechanisms: Associations are established between multiple patches, and through interaction mechanisms, the model can pass information from one patch to another. Combining self-attention mechanisms to capture internal sequence dependencies. Adaptive mixed convolutional layers can be used to learn the correlations between different patches at each time scale autonomously. Multi-feature attention: For multiple input features, attention mechanisms are designed to allow features to influence each other. For example, feature weights can be dynamically adjusted based on similarity, giving dominant features higher priority in modeling.

Through the aforementioned multi-scale time–frequency analysis theories and methods, the life data of power electronics can be effectively analyzed and predicted, thereby enhancing the reliability and performance of equipment. These methods provide important technical support for the intelligent management and maintenance of power supplies.

8.4.2 The Theoretical Basis of Preformer

The Preformer [3] is a Transformer-based model specifically designed for long-term time series forecasting. It primarily introduces the Multi-Scale Segment Correlation (MSSC) mechanism and the Predictive Multi-Scale Segment Correlation (PreMSSC) paradigm. The main structure of the model is depicted in Fig. 8.8.

The encoder is made up of N identical layers, each comprising MSSC module and a feed-forward network, followed by a sequence decomposition module with residual connections. The input to the encoder is $X_1{:}t_0$, representing all time series values from the first time point (t_1) to the current time point (t_0), along with time-related covariates. The sequence decomposition module breaks down the input into seasonal and trend components using an average pooling layer. All decomposition modules in the encoder remove the trend component, enabling the encoder to focus on modeling seasonal patterns.

The decoder consists of M identical layers and includes additional PreMSSC modules, where the key and value matrices are derived from the encoder outputs at each decoder layer. The decoder's input consists of two parts: the seasonal component and the trend component. The decomposition module in the decoder progressively extracts the trend component from the variables and adds it to the seasonal component to generate the decoder output. A fully layer takes the decoder's output and trend features as inputs to produce the final prediction.

The core mechanism of the model is the MSSC mechanism. Each segment is a subsequence of the time series, containing local information of the time series at a certain time scale. The process of segment division begins by determining the segment length $Lseg$, a hyperparameter that decides the number of time points in each segment. Then, the time series $x \in \mathbb{R}^{L\times d}$ is divided into multiple segments of length $Lseg$. The Q, K, and V generated by x are also divided into segments of the same length: $\{Q_1, Q_2, \ldots, Q_m\}$, $\{K_1, K_2, \ldots, K_n\}$,$\{V_1, V_2, \ldots, V_n\}$,. Where, $Q_i \in \mathbb{R}^{L_{seg}\times d}$, $K_i \in \mathbb{R}^{L_{seg}\times d}$, $V_i \in \mathbb{R}^{L_{seg}\times d}$. The correlation c_{ij} between Q and K can be calculated using the following formula:

$$c_{ij} = \text{Correlation}\left(Q_i, K_j\right) = \frac{1}{d \times L_{seg}} Q_i \odot K_j \tag{8.16}$$

The correlation degree is obtained by the softmax function:

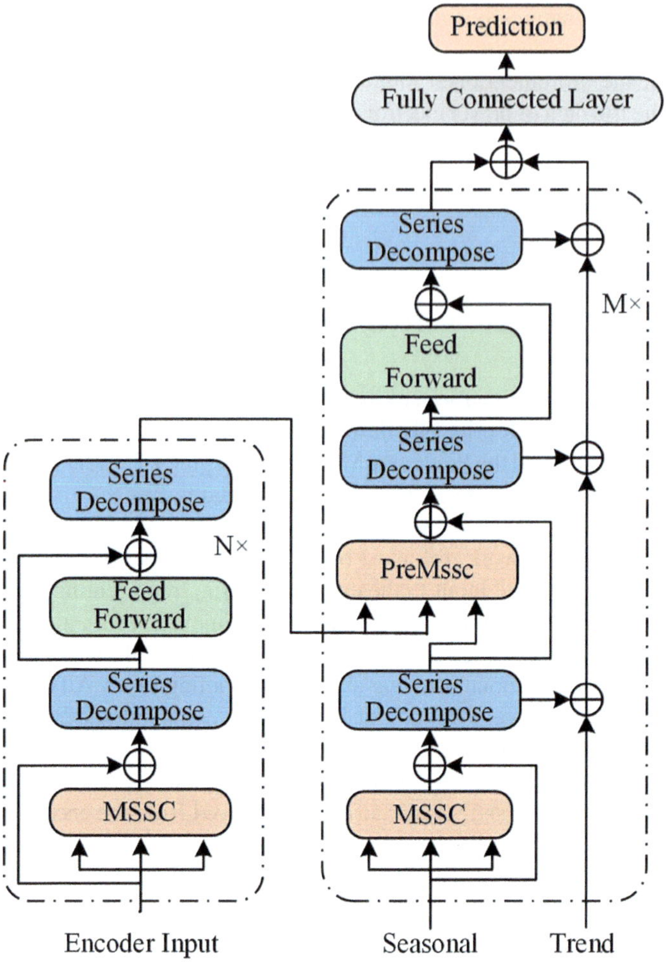

Fig. 8.8 Framework of the Preformer

$$\hat{c}_{i1},\hat{c}_{i2},\ldots,\hat{c}_{in} = \text{Softmax}(c_{i1},c_{i2},\ldots,c_{in}). \tag{8.17}$$

Then, Segment-Correlation is obtained by multiplying and V respectively

$$\begin{aligned} Y_i &= \sum_{j=1}^{n} \hat{c}_{ij} V_j \\ \text{SC}(H; L_{\text{seg}}) &= \text{Concat}(Y_1, \ldots, Y_m) \end{aligned} \tag{8.18}$$

The outputs from multiple Segment-Correlation layers with different L_{seg} are integrated in the MSSC mechanism. Specifically, with the increase of scale level, the segment length starts from a small initial value L_0 and exponentially increase the segment length, i.e., $L_l = 2^l L_0$, where $l \in \{0, 1, \ldots, l_{\max}\}$ represents the scale level, $l_{\max} = \left\lceil \log_2\left(\frac{L}{L_0}\right)\right\rceil$. The input H is passed through Segment-Correlation layers at various scale levels to obtain the corresponding outputs. These outputs are then aggregated across all scale levels to produce the final output of the MSSC module.

$$\mathrm{MSSC}(H) = \sum_{l=0}^{l_{\max}} \alpha_l \cdot \mathrm{SC}\left(H; 2^l L_0\right),$$
$$\alpha_l = \frac{2^l}{\sum_{l=0}^{l_{\max}} 2^l} \tag{8.19}$$

8.4.3 Experimental Result and Analysis

In this section, the Crossforme [25], Preformer, Pyraformer [9], PatchTST [13], and Scaleformer [19] models are used to predict the remaining useful life of range-extended electric vehicle power batteries under two different test conditions: CLTC and NEDC. The results of the models are presented in Table 8.6 and Fig. 8.9. Upon comparison of the five models, it is evident that the Preformer exhibits superior overall performance. The detailed experimental analysis is presented as follows:

All five models employ different methods to enhance multi-scale characteristics, but the Preformer is more suitable for battery life prediction. For instance, the Preformer model achieved a score of 0.1923 under the NEDC condition, which is an improvement of 56.97–29.71% compared to other models. Although the Pyraformer

Table 8.6 Comparison of predictive metrics based on Preformer

Model	Dataset	MAE	MSE	MAPE	R^2
Crossformer	CLTC	0.3266	0.3469	1.3584	0.7019
	NEDC	0.3865	0.3784	1.6523	0.7433
Preformer	CLTC	0.2009	0.1926	0.5749	0.8720
	NEDC	0.1923	0.1897	0.6198	0.8941
Pyraformer	CLTC	0.4264	0.4631	2.3421	0.6421
	NEDC	0.4469	0.4812	2.6481	0.6797
PatchTST	CLTC	0.2463	0.2689	0.9164	0.7930
	NEDC	0.2736	0.2543	0.8611	0.7213
Scaleformer	CLTC	0.3863	0.4083	1.2145	0.6531
	NEDC	0.3236	0.3941	1.2477	0.7027

Fig. 8.9 Comparison of prediction results based on Preformer

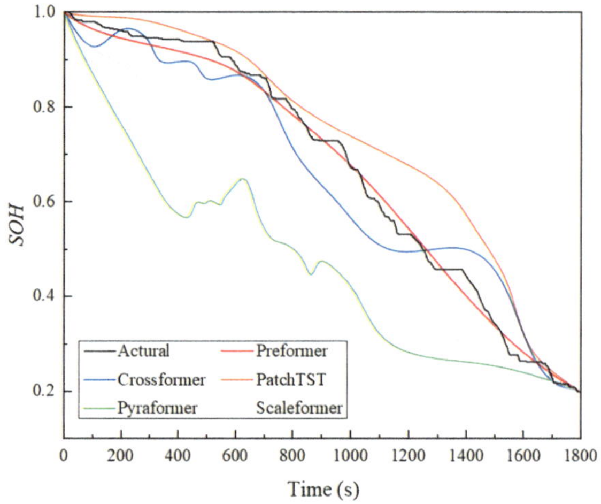

(a) The CLTC condition

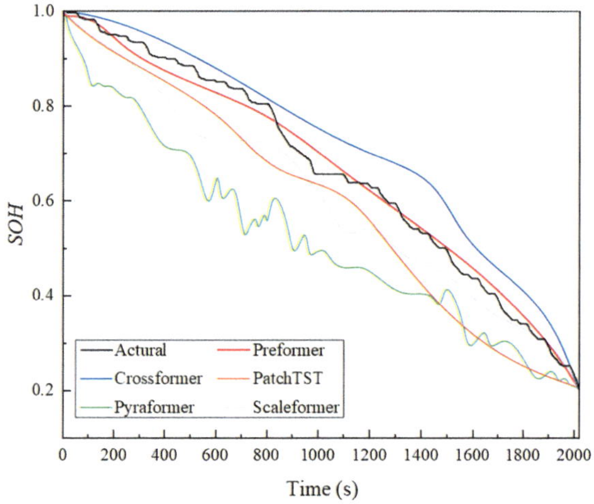

(b) The NEDC condition

and Scaleformer also use modified attention mechanisms to enhance feature extraction capabilities, the Preformer's multi-scale segment correlation mechanism can accurately capture the intrinsic features.

Compared to the Patch preprocessing mechanism used by PatchTST, the Preformer's improvement. For example, in the MSE comparison under CLTC and NEDC conditions, the models with better performance include PatchTST and Preformer. Although the Patch mechanism has become a necessary module in recent long-term forecasting model designs, such as Crossformer which also uses patches to exchange feature relationships between time and variables, the Preformer's combination of sequence decomposition, multi-component processing, and multi-scale segment correlation is more capable of making accurate predictions.

8.5 Model Performance Comparison

In this chapter, ten deep learning models include Autoformer, FEDformer, Informer, N-Transformer, Transformer, Crossformer, Preformer, Pyraformer, PatchTST, and Scaleformer were employed to test the predictive performance on the remaining lifespan of range-extended electric vehicle power batteries. Two test conditions for power batteries, CLTC and NEDC, were selected for analysis. This chapter explores the potential of Transformer models in life prediction tasks and discusses the application differences of multi-scale enhancement. Based on comparative experiments of predictive performance, the performance comparison among various metrics is shown in Fig. 8.10, and a summary of the model performances is presented as follows:

Transformer-based forecasting algorithms perform well in predicting the remaining useful life of power batteries, with metrics generally meeting forecasting requirements. This is mainly due to several advantages of Transformers. Firstly, the self-attention mechanism can directly compute the relationship between any two-time steps in a sequence, a flexibility that allows the model to effectively capture long-distance dependencies in time series. In contrast, traditional time series models (such as ARIMA or RNN) often struggle to capture dependencies over long time spans. Secondly, Transformers can consider the entire input sequence when processing each time step, thereby obtaining a more comprehensive context. This is particularly important for data lacking periodicity or trends, as the model can make predictions by recognizing complex and random patterns. Lastly, the multi-head attention mechanism enables the model to learn different dependencies at the same time. For instance, in situations where multiple factors influence temporal changes, the model can simultaneously monitor the changes in several important variables, thereby improving the accuracy of predictions.

Models that enhance the multi-head attention mechanism in a comprehensive manner tend to perform better. For instance, FEDformer decouples the length of the input sequence from the dimension of the attention matrix by leveraging frequency domain enhancement. Meanwhile, Preformer incorporates a multi-scale segment correlation attention mechanism. This is mainly because multi-perspective learning enables the model to capture a variety of relationships and patterns between different positions or features within the sequence. The enhanced multi-head attention mechanism can selectively focus on key parts of the input based on the specific task requirements. This adaptability allows the model to perform better across different contexts and tasks. Additionally, some improvements, such as sparse attention, can reduce computational time and memory usage, which is particularly crucial for processing long sequence data. Traditional full attention mechanisms often face significant computational bottlenecks when dealing with long sequences.

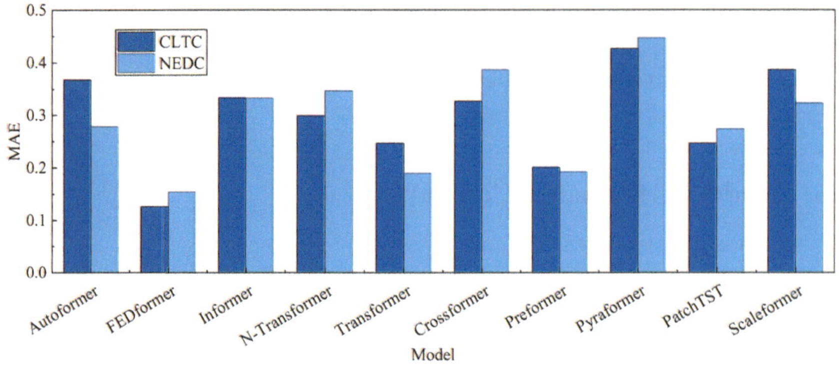

(a) MAE

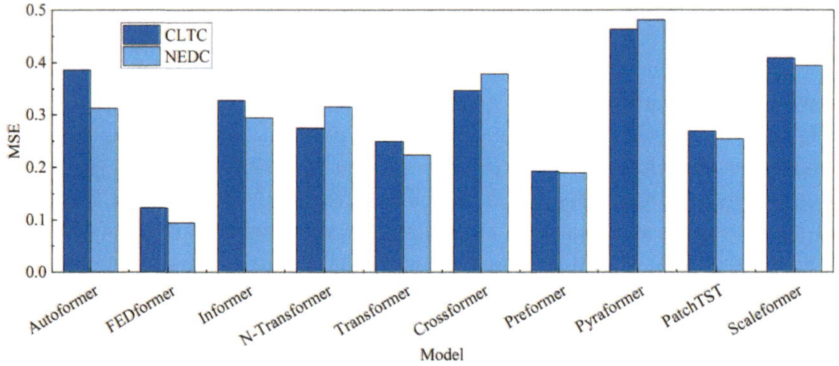

(b) MSE

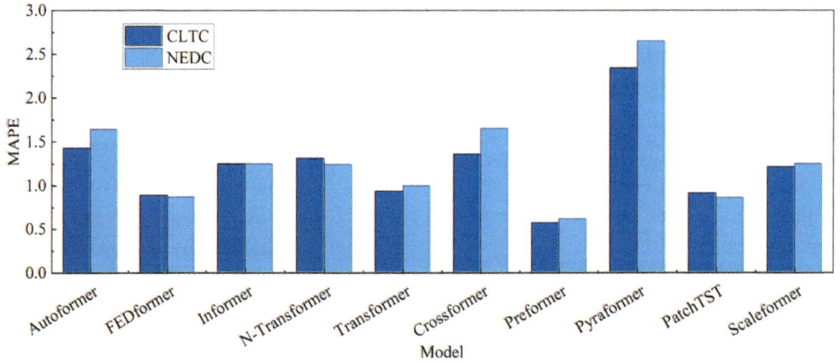

(c) MAPE

Fig. 8.10 Model performance comparison of all metrics

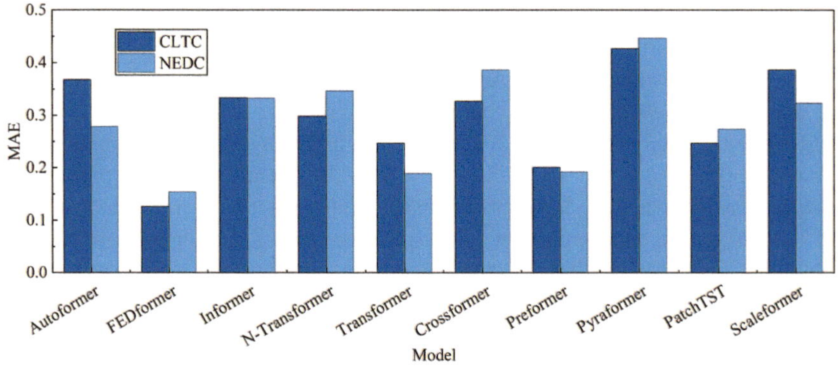

(d) R^2

Fig. 8.10 (continued)

The application of multi-scale methods in battery life prediction does not show a significant effect compared to the original Transformer model. This is mainly because the methods used by these multi-scale models for processing the seasonal and trend components of time series data are not suitable for the battery data used. The battery data uses two test conditions, CLTC and NEDC, which are more pronounced in trend terms but do not exhibit seasonality. Multi-scale methods usually require optimization for specific features, meaning that if the features of the input data do not show significant differences across different scales, the benefits of using multi-scale methods may be limited. In battery life prediction, commonly used features may already be fully captured by the Transformer model. On the other hand, multi-scale methods usually require optimization for specific features, meaning that if the features of the input data do not show significant differences across different scales, the benefits of using multi-scale methods may be limited. In battery life prediction, commonly used features may already be fully captured by the Transformer model.

8.6 Conclusions

In recent years, battery RUL prediction has made significant progress. Current research focuses mainly on optimizing algorithm architectures, multi-scale fusion methods, and developing more performant lightweight models. However, to advance the field, several areas require further enhancement:

Enhancing RUL prediction accuracy and early prediction capabilities. To improve the accuracy of RUL predictions, researchers need to construct comprehensive and high-quality datasets that include various sensor data throughout the entire lifecycle of the battery (such as charging and discharging current, voltage, temperature, etc.). Prediction models should be built using a variety of machine learning algorithms, and model parameters should be optimized through methods. Early prediction capabilities are an essential part of battery management systems, enabling fault detection and reducing resource consumption. Additionally, researchers should focus on how to use limited initial cycle data for accurate RUL predictions.

Modeling capacity regeneration. Capacity regeneration is a common phenomenon in the battery aging process and directly affects the accuracy of RUL predictions. Studies have shown that depth of discharge and high-temperature environments can exacerbate the capacity regeneration phenomenon. Therefore, future research should aim to enable machine learning algorithms to learn and recognize the characteristics of capacity regeneration, especially in predictions based on SOH estimation. Researchers can explore dynamically adjusting machine learning strategies throughout the entire aging cycle, selecting the most appropriate algorithm based on different capacity regeneration intervals, thereby improving the reliability of RUL predictions.

Integrating physical principles. The deep integration of physical principles with machine learning algorithms has shown many advantages, including improving prediction accuracy. Current research has successfully incorporated physical principles into neural network models, and future exploration should further investigate diverse ways to combine physical principles with ML algorithms. For example, when using kernel functions based on Arrhenius's law, adjusting the kernel function to include physical factors such as temperature, depth of discharge, and regenerated capacity in polynomial equations or periodic covariance functions within the covariance function can achieve more accurate modeling of complex physical processes.

References

1. Cai ST, Hu J, Ma SQ, Yang ZN, Wu H (2022) Remaining useful life prediction method of EV power battery for DC fast charging condition. Energy Rep 8:1003–1010. https://doi.org/10.1016/j.egyr.2022.08.095
2. Chou JH, Wang FK, Lo SC (2023) Predicting future capacity of lithium-ion batteries using transfer learning method. J Energy Storage 71:108120. https://doi.org/10.1016/j.est.2023.108120
3. Du DZ, Su B, Wei ZW (2023) Preformer: predictive transformer with multi-scale segment-wise correlations for long-term time series forecasting. In: ICASSP 2023–2023 IEEE international conference on acoustics, speech and signal processing (ICASSP). IEEE, pp 1–5
4. Hasib SA, Islam S, Chakrabortty RK, Ryan MJ, Saha DK, Ahamed MH, Moyeen SI, Das SK, Ali MF, Islam MR, Tasneem Z, Badal FR (2021) A comprehensive review of available battery datasets, RUL prediction approaches, and advanced battery management. IEEE Access 9:86166–86193. https://doi.org/10.1109/ACCESS.2021.3089032
5. Hossain LMS, Ansari S, Miah MS, Meraj ST, Hasan K, Shihavuddin ASM, Hannan MA, Muttaqi KM, Hussain A (2022) Deep learning enabled state of charge, state of health and remaining useful life estimation for smart battery management system: methods, implementations, issues and prospects. J Energy Storage 55:105752. https://doi.org/10.1016/j.est.2022.105752

6. Jafari S, Byun YC (2023) Optimizing battery RUL prediction of lithium-ion batteries based on Harris hawk optimization approach using random forest and LightGBM. IEEE Access 11:87034–87046. https://doi.org/10.1109/ACCESS.2023.3304699
7. Jafari S, Byun YC (2024) Accurate remaining useful life estimation of lithium-ion batteries in electric vehicles based on a measurable feature-based approach with explainable AI. J Supercomput 80(4):4707–4732. https://doi.org/10.1007/s11227-023-05648-8
8. Li XY, Shu X, Shen JW, Xiao RX, Yan WS, Chen Z (2017) An on-board remaining useful life estimation algorithm for lithium-ion batteries of electric vehicles. Energies 10(5):691. https://doi.org/10.3390/en10050691
9. Liu SZ, Yu H, Liao C, Li JG, Lin WY, Liu AX (2022) Dustdar Schahram Pyraformer: low-complexity pyramidal attention for long-range time series modeling and forecasting
10. Liu Y, Wu HX, Wang JM, Long MS (2022) Non-stationary transformers: exploring the stationarity in time series forecasting. Adv Neural Inf Process Syst 35:9881–9893
11. Ly S, Xie JH, Wolter F, Nguyen HD, Weng Y (2023) T-shape data and probabilistic remaining useful life prediction for Li-ion batteries using multiple non-crossing quantile long short-term memory. Appl Energy 349:121355. https://doi.org/10.1016/j.apenergy.2023.121355
12. Meng FB, Yang FF, Yang J, Xie M (2023) A power model considering initial battery state for remaining useful life prediction of lithium-ion batteries. Reliab Eng Syst Saf 237:109361. https://doi.org/10.1016/j.ress.2023.109361
13. Nie YQ, Nguyen NH, Sinthong P, Kalagnanam J (2023) A time series is worth 64 words: long-term forecasting with transformers. https://doi.org/10.48550/arXiv.2211.14730
14. Pang XQ, Huang R, Wen J, Shi YH, Jia JF, Zeng JC (2019) A lithium-ion battery RUL prediction method considering the capacity regeneration phenomenon. Energies 12(12):2247. https://doi.org/10.3390/en12122247
15. Qi DF, Li CB, Wang NB, Huang ML, Hu ZM, Li W (2021) A novel approach investigating the remaining useful life predication of retired power lithium-ion batteries using genetic programming method. J Electrochem Energy Convers Storage 18(3):030904. https://doi.org/10.1115/1.4050510
16. Rincón-Maya C, Guevara-Carazas F, Hernández-Barajas F, Patino-Rodriguez C, Usuga-Manco O (2023) Remaining useful life prediction of lithium-ion battery using ICC-CNN-LSTM methodology. Energies 16(20):7081. https://doi.org/10.3390/en16207081
17. Saleem U, Liu WJ, Riaz S, Li WL, Hussain GA, Rashid Z, Arfeen ZA (2023) TransRUL: a transformer-based multihead attention model for enhanced prediction of battery remaining useful life. Energies 17(16):3976. https://doi.org/10.3390/en17163976
18. Samanta A, Williamson S (2023) Machine learning-based remaining useful life prediction techniques for lithium-ion battery management systems: a comprehensive review. IEEJ J IA 12(4):563–574. https://doi.org/10.1541/ieejjia.22004793
19. Shabani A, Abdi A, Meng LL, Sylvain T (2023) Scaleformer: iterative multi-scale refining transformers for time series forecasting. https://doi.org/10.48550/arXiv.2206.04038
20. Shi JC, Rivera A, Wu DZ (2022) Battery health management using physics-informed machine learning: online degradation modeling and remaining useful life prediction. Mech Syst Signal Process 179:109347. https://doi.org/10.1016/j.ymssp.2022.109347
21. Wang FJ, Zhao ZB, Ren JX, Zhai Z, Wang SB, Chen XF (2022) A transferable lithium-ion battery remaining useful life prediction method from cycle-consistency of degradation trend. J Power Sources 521:230975. https://doi.org/10.1016/j.jpowsour.2022.230975
22. Wu HX, Xu JH, Wang JM, Long MS (2021) Autoformer: decomposition transformers with auto-correlation for long-term series forecasting. Adv Neural Inf Process Syst 34:22419–22430
23. Yan T, Chen JZ, Hui D, Li XJ, Zhang DL (2024) The remaining useful life forecasting method of energy storage batteries using empirical mode decomposition to correct the forecasting error of the long short-term memory model. Sustainability 16(5):1920. https://doi.org/10.3390/su16051920
24. Zhang YZ, Xiong R, He HW, Pecht MG (2018) Long short-term memory recurrent neural network for remaining useful life prediction of lithium-ion batteries. IEEE Trans Veh Technol 67(7):5695–5705. https://doi.org/10.1109/TVT.2018.2805189

25. Zhang YH, Yan JC (2023) Crossformer: transformer utilizing cross-dimension dependency for multivariate time series forecasting. In: The eleventh international conference on learning representations
26. Zhou HY, Zhang SH, Peng JQ, Zhang S, Li JX, Xiong H, Zhang WC (2021) Informer: beyond efficient transformer for long sequence time-series forecasting. In: Proceedings of the AAAI conference on artificial intelligence, vol 12, pp 11106–11115
27. Zhou T, Ma ZQ, Wen QS, Wang X, Sun L, Jin R (2022) Fedformer: frequency enhanced decomposed transformer for long-term series forecasting. In: International conference on machine learning. PMLR, pp 27268–27286

Chapter 9
Big Data Embedding in PHM for Electromechanical System

9.1 Introduction

As industries worldwide gradually move towards intelligent and digital transformations, the application scope of intelligent electromechanical systems is expanding, encompassing areas from industrial manufacturing to medical devices, from transportation to energy management, and even to home automation. The complexity and critical nature of these systems make ensuring their long-term stable operation a pressing issue that needs to be addressed. In this context, PHM technology has emerged and rapidly developed into one of the core technologies for ensuring the healthy operation of intelligent electromechanical systems [7]. PHM monitors and analyzes operational data from equipment to predict potential failures in advance and formulate reasonable maintenance plans, thereby effectively reducing unexpected downtime and maintenance costs, and extending the lifespan of equipment [8].

However, as intelligent electromechanical systems are increasingly deployed, the volume of data generated by these systems is growing exponentially. This data includes not only conventional mechanical vibration signals and electrical parameters but also environmental factors, operational history, and fault records across multiple dimensions. The enormous scale and diversity of this data present significant challenges to traditional PHM methods. Traditional approaches typically rely on limited data and predefined physical models, which often exhibit inefficiencies and insufficient predictive accuracy when faced with today's complex and variable real-world conditions and vast amounts of data. Therefore, effectively handling and utilizing big data within PHM systems has become a current research focus.

Against this backdrop, big data technology [11] is widely regarded as key to enhancing PHM system performance. Big data technology enables the processing and storage of large volumes of data while also uncovering valuable insights and patterns through techniques like deep learning and data mining. This enhances the predictive accuracy and real-time capabilities of PHM. Modern big data processing technologies like Hadoop [17], Spark [23], and Flink [21] provide powerful distributed computing

H. Liu et al., *Prognostics and Health Management for Intelligent Electromechanical Systems*, https://doi.org/10.1007/978-981-96-7218-9_9

capabilities, enabling efficient and precise processing of petabyte-scale data. Additionally, new data management technologies, such as NoSQL databases and cloud storage, provide a solid foundation for the large-scale data storage and management needs of PHM systems.

Reflecting on the earlier chapters of this book, we have thoroughly discussed fault diagnosis and life prediction models for various critical components in intelligent electromechanical systems [13], such as bearings [5], IGBTs [1], MOSFETs, capacitors, and power supplies. These models have achieved remarkable results on small-scale datasets, but as data scales increase and application scenarios become more complex, traditional standalone models and data processing methods are clearly insufficient to meet the challenges of the new landscape. Particularly in fields such as industrial robots, CNC machines, drones, Da Vinci surgical robots, autonomous vehicles, and smart grids, the operational data is not only massive but also complex, imposing higher demands on the timeliness, accuracy, and reliability of PHM systems [22, 30].

Therefore, the core objective of this chapter is to explore how to embed big data technology into PHM systems to significantly enhance their performance and adaptability. We will first explore the construction of big data storage platforms, analyzing the data storage and management needs of intelligent electromechanical systems in detail, including how to ensure data scalability, reliability, and real-time access. Specifically, we will introduce how to use Hadoop Distributed File System (HDFS) [22, 30], NoSQL databases [32], and cloud storage solutions [4] to build an efficient storage platform that supports the persistence and rapid retrieval of large-scale data.

Next, we will delve into the development of distributed predictive models to fully leverage the computing resources available in a big data environment, enabling accurate predictions of the operational status of intelligent electromechanical systems. In this section, we will introduce several mainstream distributed computing frameworks, including Hadoop, Apache Spark, and Flink, and discuss their application scenarios and advantages in processing large-scale data. By applying these technologies, PHM systems can achieve full-process optimization from data acquisition and storage to analysis and prediction, thus continuing to provide efficient and accurate maintenance recommendations even when faced with complex real-world conditions.

However, the application of big data in PHM systems is not without challenges. In practice, we may encounter issues such as data heterogeneity, high-dimensional data processing, and the need for real-time data handling. This chapter will discuss these challenges in detail and propose corresponding solutions, such as using data fusion techniques to address heterogeneous data integration, employing dimensionality reduction and feature selection to handle data, and enhancing system real-time capabilities by combining edge computing with cloud computing.

Finally, this chapter will summarize the broad impact of embedding big data technology into PHM systems and look ahead to future development directions. With the development of the Industrial IoT [20, 26, 31] and intelligent manufacturing, the intelligence and autonomy of PHM systems will become a trend. The discussions in this chapter not only provide strong support for the current PHM of intelligent electromechanical systems but also lay a foundation for the future development of Industry 4.0 and intelligent industries.

9.2 Construction of Big Data Storage Platform

With the widespread application of intelligent electromechanical systems across various fields, the amount of data generated during system operation is experiencing explosive growth. This data includes, but is not limited to, sensor data, operation logs, environmental monitoring data, operation records, and fault reports. Given the vast scale of such data, traditional methods of data storage and management can no longer effectively meet current demands. As a result, building an efficient, scalable, and reliable big data storage platform has become a key element in the PHM system for intelligent electromechanical systems [27, 29].

This big data storage platform must not only possess strong data processing capabilities but also handle the heterogeneity of different data types and meet the real-time processing needs of massive datasets. By integrating advanced cloud computing technologies, distributed storage systems, and big data analytics tools, PHM systems for intelligent electromechanical systems can achieve real-time monitoring and accurate prediction of equipment operating conditions, thereby improving system reliability, enhancing maintenance efficiency, reducing failure rates, and lowering maintenance costs.

In constructing such a platform, it is also crucial to consider data security and privacy protection to ensure that the critical data generated during system operation is stored and safeguarded effectively. Additionally, the platform should offer flexible scalability to accommodate future data growth and meet the long-term development needs of intelligent electromechanical systems. Figure 9.1 shows the overall hierarchical structure of the big data ecosystem from data storage, resource scheduling, and computing framework to data processing applications.

9.2.1 Data Source and Acquisition

The big data in intelligent electromechanical systems is characterized by diversity and complexity, originating from various devices and sensors and covering a wide range of physical quantities. To effectively analyze and predict the health status of the equipment, this data must be systematically collected, stored, and preprocessed. In this section, we will present a table that illustrates the sources and characteristics of common electromechanical system PHM big datasets discussed in the previous chapters. Details are given in Table 9.1.

In a big data environment, data acquisition often involves the collaborative work of multiple sensors. For each data type, the first challenges to address are data loading, cleaning, and preprocessing. The data loading process involves importing

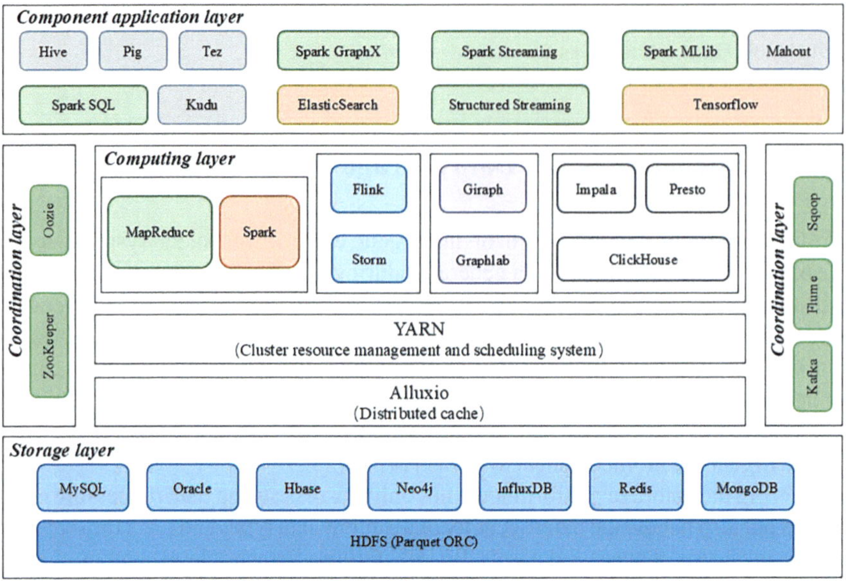

Fig. 9.1 Hierarchical architecture of big data ecosystem

Table 9.1 PHM big data set of electromechanical system

Data type	Dataset	Data scale	Data source
Bearing vibration signal	CWRU	120,000 * 4	[24]
Bearing fault signal		120,000 * 10	
Bearing remaining service life signal	PHM 2012	(120–2803) * 17	[16]
IGBT remaining service life signal	Thermal Overstress Aging with Square Signal at gate and SMU data	125,000 * 7	[25]
MOSFET remaining service life signal	MOSFET Thermal Overstress Aging	129,500 * 6	[3]
Capacitor remaining service life signal	Capacitor Electrical Stress-2	194 * 6	[2]
Power supply remaining service life signal	Extended range electric vehicle battery life	19,993 * 16 (CLTC standard)	Vehicle condition test simulation (based on Simulink and Cruise)
		22,663 * 16 (NEDC standard)	

raw data from various sources into the system and performing initial format conversions to meet the requirements for subsequent processing. The cleaning step includes removing noise from the data, handling missing values, and correcting abnormal data. Finally, the feature extraction step transforms the raw data into more representative features, facilitating further analysis and modeling.

9.2.2 Data Storage and Management Technology

Intelligent electromechanical systems involve complex multi-source data, including real-time data from sensors, equipment operation logs, fault records, historical maintenance data, and operating environment data. This data comprises not only structured data (such as time-series data in equipment operation logs) but also semi-structured and unstructured data (such as sensor signals and video surveillance data). As the scale of intelligent electromechanical systems expands, the demand for data storage and management becomes increasingly complex. Traditional relational databases are gradually showing limitations in terms of scalability and performance when faced with large-scale data storage requirements.

Scalability: The data volume in intelligent electromechanical systems is growing exponentially, especially as the number of devices increases and sensor technology advances, leading to a substantial increase in data generation speed. The system requires a storage platform capable of scaling storage and computing resources as the data volume grows. Horizontal scaling (adding more storage nodes) is key to addressing this challenge.

Reliability: Intelligent electromechanical systems are often applied in high-risk or critical fields (such as industrial production, transportation, medical equipment, etc.), making reliable data storage and backup crucial. Any data loss or storage failure can severely impact the normal operation of the system. Data redundancy and backup mechanisms are needed to ensure that data is not lost during storage.

Real-time access: PHM systems need to monitor equipment status in real-time and make predictive maintenance decisions quickly based on the data. This requires the data storage system to have low-latency read/write capabilities to ensure that data can be stored and accessed by the analysis system in a timely manner.

Data persistence and retrieval: In intelligent electromechanical systems, certain data needs to be stored for long periods, particularly historical operation and fault data, which are valuable for future equipment health management and trend prediction. An efficient storage system must not only ensure data persistence but also support fast retrieval capabilities to handle complex queries.

To meet the big data storage requirements of PHM systems in intelligent electromechanical systems, this section will introduce several commonly used big data storage technologies, including the HDFS, NoSQL databases, and cloud storage solutions.

(1) *Hadoop distributed file system*

The HDFS [28] is a highly scalable and fault-tolerant file system, specifically designed for storing and processing large-scale unstructured data. The core concept of HDFS is to utilize distributed storage and computing to spread large amounts of data across multiple nodes, while ensuring data security and availability through redundancy and replication mechanisms. In PHM systems, HDFS can be employed to store massive sensor data, log files, and historical operation records, which often exhibit a high degree of redundancy and distributed characteristics.

- Architecture Overview: HDFS employs a master–slave architecture, where a single NameNode acts as the master, and multiple DataNodes serve as the slaves. The NameNode is responsible for managing the file system's metadata, such as the file directory structure and the mapping between files and data blocks. DataNodes are responsible for the actual storage of data blocks and handling read/write requests from clients.
- Data Block and Storage: In HDFS, files are divided into equally sized data blocks (with a default block size of 128 MB), which are distributed across multiple DataNodes. Each data block is typically replicated across multiple nodes (with a default replication factor of 3) to provide data redundancy and fault tolerance.
- Data Read/Write Process: When a client writes data to HDFS, the data is first split into blocks, which are then written sequentially to multiple DataNodes in a pipeline manner. The NameNode maintains the mapping information for these data blocks so that files can be correctly reassembled during read operations. For read requests, the NameNode locates the blocks of a file, and the client retrieves the data directly from the relevant DataNodes.
- Fault Tolerance Mechanism: HDFS has strong fault-tolerant capabilities, mainly achieved through the replication mechanism of data blocks. If a DataNode fails, the NameNode detects the failure immediately and replicates the affected data blocks to other healthy DataNodes to ensure data availability.
- Application Scenarios: HDFS is particularly suitable for scenarios that require the storage and processing of large-scale unstructured data, such as log files, sensor data, and image/video files. Due to its design for batch-processing operations, HDFS performs exceptionally well in batch-processing applications that require high throughput for data access.
- *NoSQL database*

NoSQL databases are designed to meet the data management demands in big data environments, capable of handling high concurrency, large-scale data storage, and flexible data models. Popular NoSQL databases such as Cassandra, MongoDB, and HBase are widely used in PHM systems. These databases not only support horizontal scalability and efficient data read/write operations but also manage various data types, including key-value pairs, documents, column families, and graph data. This makes NoSQL databases particularly suitable for storing and managing data from different sensors and devices, especially when such data is unstructured or semi-structured.

- Key-Value Stores: It is a fundamental NoSQL database model, where data is saved as pairs of keys and associated values, with each key being distinct. Redis and Riak are typical key-value store databases. Key-value stores are well-suited for scenarios requiring fast lookup and writing, such as caching and session management.

- Document Stores: Document store databases store data in the form of documents, typically represented in formats like JSON, BSON, or XML. MongoDB is a leading document store database, allowing nested documents and array structures, which can naturally represent complex data relationships. Document stores are ideal for handling semi-structured data, such as log records, user profiles, and content management system data.
- Column Family Stores: Column family store databases (such as Apache Cassandra and HBase) are based on Google's Bigtable design, storing data by column families, with each family containing multiple related columns. Column family stores are particularly suitable for handling large-scale time-series data and multidimensional datasets, such as sensor data and user behavior data.
- Graph Databases: Graph databases, such as Neo4j and Amazon Neptune, store data in the form of nodes and edges, making them especially suited for managing complex relational data, such as social networks, supply chains, and recommendation systems.
- Consistency Models: NoSQL databases typically follow the CAP theorem, which involves trade-offs between consistency, availability, and partition tolerance. Different NoSQL databases prioritize these three aspects differently. For example, MongoDB emphasizes availability and partition tolerance, while Cassandra compromises between availability and partition tolerance.
- Application Scenarios: NoSQL databases are widely used in scenarios requiring flexible data models and high concurrency access, such as product catalogs for e-commerce websites, IoT sensor data storage, and social media data analytics.
- *Cloud storage solutions*

Cloud storage has become one of the mainstream choices for big data storage. Cloud storage is a service provided over a network, based on cloud computing architecture, allowing users to store and access data on demand. Cloud storage solutions typically consist of multiple distributed data centers, offering high availability and scalability. In PHM systems, cloud storage can serve as the primary data repository or as a backup solution for local storage, ensuring the security and accessibility of the data.

- Types of Storage:
  - Object Storage: Object storage (such as Amazon S3, Google Cloud Storage [19]) is one of the primary forms of cloud storage, where data is stored as objects. Each object includes the data, metadata, and a unique ID. Object storage is well-suited for handling large volumes of unstructured data like backups, logs, images, and videos
  - Block Storage: Block storage (such as Amazon EBS, Google Persistent Disk) provides storage services similar to traditional hard drives, making it suitable for data storage scenarios that require low latency and high performance. Block

storage is typically used for disk storage in virtual machines and database storage.
  - File Storage: File storage (such as Amazon EFS, Google Filestore) offers a traditional file system interface, supporting folders, directory structures, and permission management. It is suitable for shared storage and migrating traditional applications to the cloud.
  - Data Distribution and Redundancy: Cloud storage typically spreads data across several data centers in different locations to guarantee high availability and support disaster recovery. Through data replication and redundant storage, cloud storage can maintain data accessibility even in the event of hardware failures or natural disasters.

- Access Control and Security: Cloud storage provides fine-grained access control policies, allowing users to set different levels of access permissions. Additionally, cloud storage services generally support data encryption (both at rest and in transit) to ensure data security. Most cloud storage providers also adhere to industry security standards, such as ISO 27001 and SOC 2.
- Integration and Scalability: Cloud storage solutions are usually tightly integrated with cloud computing platforms. For example, AWS cloud storage seamlessly integrates with its computing services (such as EC2 and Lambda), enabling users to easily leverage data stored in the cloud for computing, analysis, and machine learning tasks. Furthermore, the elastic scalability of cloud storage allows users to dynamically adjust storage capacity and performance according to business needs.
- Application Scenarios: Cloud storage is widely used in various scenarios, including large-scale data storage and backup, media content distribution, enterprise-level data lakes, disaster recovery, and real-time data analytics.

HDFS, NoSQL databases, and cloud storage solutions each possess unique technical characteristics that meet the diverse big data storage requirements of PHM systems in intelligent electromechanical systems. HDFS offers batch processing and data persistence for large-scale data, making it well-suited for storing historical data and supporting offline analysis tasks. NoSQL databases provide flexible data models and high concurrency handling capabilities, making them ideal for real-time data storage and retrieval. Cloud storage solutions, with their elastic scalability, global distribution, and high reliability, are perfect for long-term data storage and disaster recovery needs in intelligent electromechanical systems. By integrating these technologies, PHM systems can more effectively tackle the challenges of large-scale data processing, ensuring efficient system operation and accurate predictive maintenance.

(4) *Data management and organization*

In addition to selecting appropriate storage technologies, efficiently organizing and managing data is key to building a robust big data storage platform. The core of data management is ensuring data availability, integrity, and security throughout its entire lifecycle. Common data management techniques include data partitioning, indexing, compression, and Data Lifecycle Management (DLM) [15].

- Data Partitioning: This technique involves dividing large datasets into smaller, independent parts based on specific rules. Partitioning improves data retrieval efficiency and reduces the amount of data processed in a single query.
- Indexing: By creating indexes for the data, retrieval speeds can be significantly increased, especially for frequently accessed or queried data. Indexing allows the system to locate specific pieces of data more quickly.
- Compression: One of the challenges of large-scale data storage is the continuous growth of data. Data compression techniques can reduce storage space consumption without losing information, making it an essential tool for managing storage capacity.
- Data Lifecycle Management: DLM defines the entire lifecycle of data from its creation to archival. It ensures that appropriate storage strategies are applied at each stage of the data's life, optimizing both performance and storage costs.

By selecting the right storage technologies and applying effective data management strategies, PHM systems can handle and leverage the vast amounts of data generated by intelligent electromechanical systems, enabling more efficient and accurate fault diagnosis and prediction.

In summary, building a robust big data storage platform is fundamental to realizing the capabilities of PHM in intelligent electromechanical systems. By using advanced technologies like Hadoop, NoSQL, and cloud storage, PHM systems can overcome the challenges of large-scale data storage and management, laying a solid foundation for the development of distributed predictive models.

9.3 Distributed Predictive Model for Electromechanical System

In the PHM of intelligent electromechanical systems [7], as data volume continues to grow and system complexity increases, traditional single-machine data processing and predictive models [12] gradually become inadequate in terms of real-time performance, accuracy, and scalability. To address these limitations, distributed predictive models have emerged as an important technological solution. These models leverage multiple computing nodes to process large-scale data in parallel, significantly enhancing computational efficiency and the model's ability to handle large datasets. In this section, we will explore the necessity of distributed predictive models, introduce several mainstream distributed computing frameworks, and demonstrate through a case study how to develop and deploy PHM models using these frameworks.

With the increasing number of sensors in intelligent electromechanical systems and the rising complexity of equipment operations, the amount of data generated daily has reached unprecedented levels. This data includes real-time monitoring data such as measurements of physical quantities like temperature, vibration, pressure, and stress, as well as historical maintenance records, fault logs, and operational records.

In the face of such massive data volumes, traditional single-machine processing modes present the following significant limitations:

Limited Computational Resources: The CPU, memory, and storage resources of a single-machine system are finite. When the data volume exceeds the capacity of a single machine, processing speed significantly decreases, or the system may even be unable to process all the data. This resource bottleneck prevents PHM systems from completing data analysis and fault prediction within a reasonable time frame.

Excessive Processing Time: For large-scale intelligent electromechanical systems, real-time performance is a key requirement for PHM systems. Single-machine systems often take hours or even days to complete model training and prediction tasks when dealing with large-scale data, which is unacceptable in practical applications, especially in scenarios requiring instant responses.

Poor System Scalability: Single-machine systems are not easily scalable by simply adding hardware resources to cope with the rapid growth of data volumes. In contrast, distributed systems can scale computational power by adding more nodes, accommodating the increasing data load.

Lack of Fault Tolerance: Single-machine systems lack sufficient fault tolerance when faced with hardware failures or unexpected crashes, making them prone to data loss or computation interruptions. Distributed systems, through task distribution and data redundancy, can enhance fault tolerance and system reliability.

In this context, distributed predictive models have emerged as a solution. By utilizing distributed computing, PHM systems can process large-scale data in parallel across multiple machines, significantly reducing computation time while improving the accuracy of predictions and the overall performance of the system. Figure 9.2 shows the PHM big data flow prediction architecture for electromechanical systems.

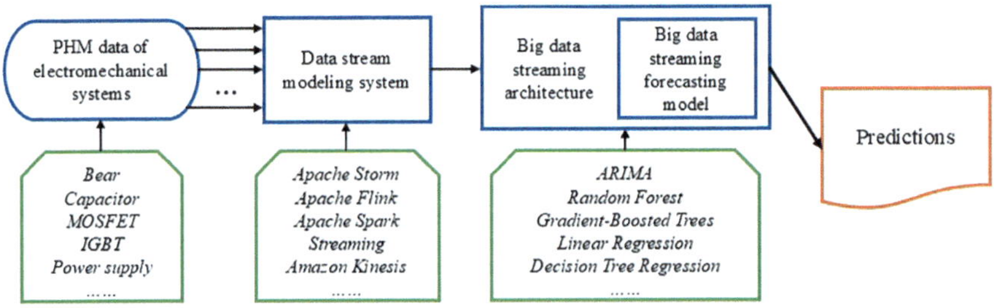

Fig. 9.2 PHM big data flow prediction architecture for electromechanical systems

9.3.1 Distributed Computing Framework

To enable the efficient development and deployment of distributed predictive models, several mainstream distributed computing frameworks are widely used in modern big data processing. This section introduces Apache Hadoop [10], Apache Spark [18], and Apache Flink [9, 21], highlighting their unique advantages in different scenarios.

(1) *Apache Hadoop*

Apache Hadoop [14] is an open-source project by the Apache Software Foundation, originally designed to process large datasets on low-cost hardware. The core components of the Hadoop ecosystem include HDFS and the MapReduce programming model.

- HDFS: As previously introduced, HDFS provides a reliable and distributed data storage layer for Hadoop. All computational tasks are processed based on data blocks stored in HDFS.
- MapReduce: MapReduce is Hadoop's core programming model. It divides tasks into multiple Map and Reduce operations. The Map tasks handle the initial processing of input data (such as filtering and sorting), while the Reduce tasks aggregate and further process the Map task outputs. By distributing the computational tasks across different nodes for parallel execution, the MapReduce model effectively handles large-scale data processing.
- Use Cases: Hadoop is well-suited for batch processing tasks, such as large-scale log analysis, historical data processing, and offline data modeling. While Hadoop performs well in batch processing, it typically has higher latency, making it unsuitable for real-time applications.

Figure 9.3 shows the basic architecture of a Hadoop cluster, including the relationship between client nodes, primary nodes, and working nodes, and the composition of the three-tier architecture (storage, resource management, and data processing).

(2) *Apache Spark*

Apache Spark offers higher computational speed and broader application scenarios compared to Hadoop. Spark's most notable feature is its in-memory computing design, which gives it a significant advantage when processing iterative algorithms and real-time data streams.

- RDD (Resilient Distributed Dataset) [6]: The core abstraction in Spark is the RDD, an immutable distributed dataset that can be operated on in parallel across a cluster. RDDs support two types of operations: transformation operations (e.g., map, filter) and action operations (e.g., reduce, collect). Through RDDs, Spark achieves large-scale parallel computation and fault tolerance.
- In-memory computing: Spark accelerates computation by caching datasets in memory, which is especially beneficial when the same dataset needs to be accessed multiple times (as in iterative machine learning algorithms). This feature allows

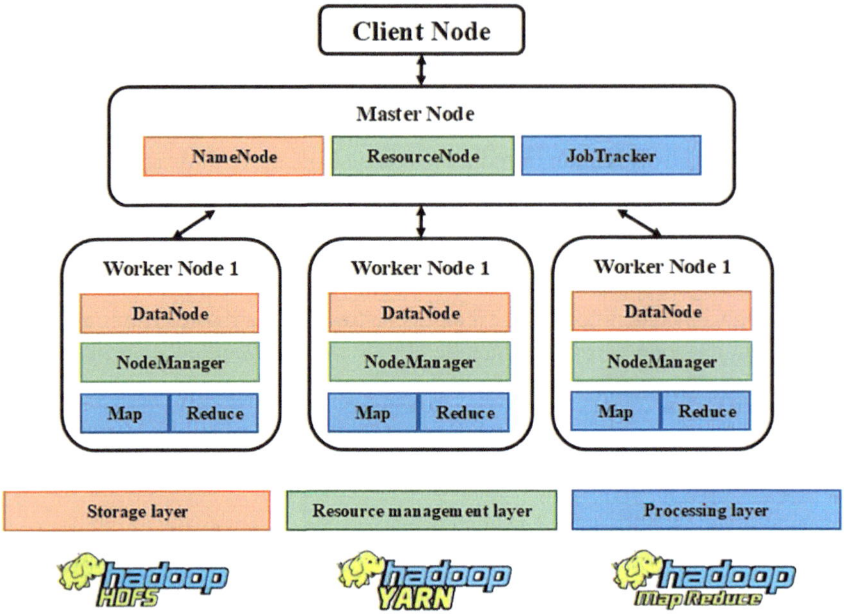

Fig. 9.3 Apache Hadoop cluster architecture diagram

Spark to perform complex computation tasks several times faster than Hadoop's MapReduce.

- Spark Streaming: Spark offers an extension module, Spark Streaming, for real-time data stream processing. By breaking real-time data into small batches and processing them using the Spark engine, Spark Streaming provides near real-time data processing capabilities.
- Use Cases: Spark is ideal for various data processing tasks, including real-time data stream processing, iterative computation, machine learning, and graph computing. It excels in high-performance, low-latency data processing scenarios, making it suitable for tasks that require fast and efficient data handling.

Figure 9.4 lists the execution process of Apache Spark, including code submission, driver task partitioning, task scheduling, and parallel data processing on working nodes.

(3) *Apache Flink*

Apache Flink is a distributed computing framework specifically designed for stream processing. Unlike Spark, Flink was built from the ground up with real-time stream processing as its core functionality, handling data streams in an event-driven manner.

- Data Stream Model: Flink employs a data stream model, treating data as a continuous stream of events. It supports stateful stream processing, meaning Flink can

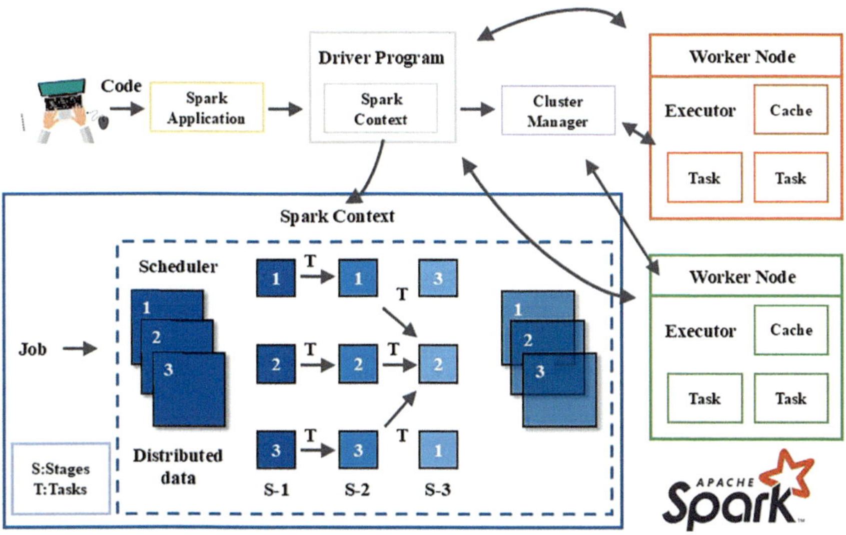

Fig. 9.4 Working principles and execution process of Apache Spark

maintain state information during the stream processing, which enables complex computations and event pattern detection.

- Low Latency and High Throughput: Flink achieves low-latency and high-throughput stream processing by splitting data streams into smaller batches for processing. Coupled with optimized memory management and network transmission mechanisms, this makes Flink excel in applications with stringent real-time requirements, such as financial transaction monitoring, IoT data processing, and real-time recommendation systems.
- Use Cases: Flink is especially suited for scenarios requiring real-time data processing and complex event handling. For example, in intelligent electromechanical systems, Flink can be used for real-time monitoring of equipment operational status, detecting anomalous events, and updating predictive models in real-time.

Figure 9.5 shows the execution architecture of Apache Flink, including the interaction between the client, JobManager, and TaskManager.

9.3.2 *Case Study*

To deepen the understanding of big data technology applications in PHM systems, a comprehensive case study is presented to validate the system's effectiveness in real-world scenarios. This study involves two experimental tasks:

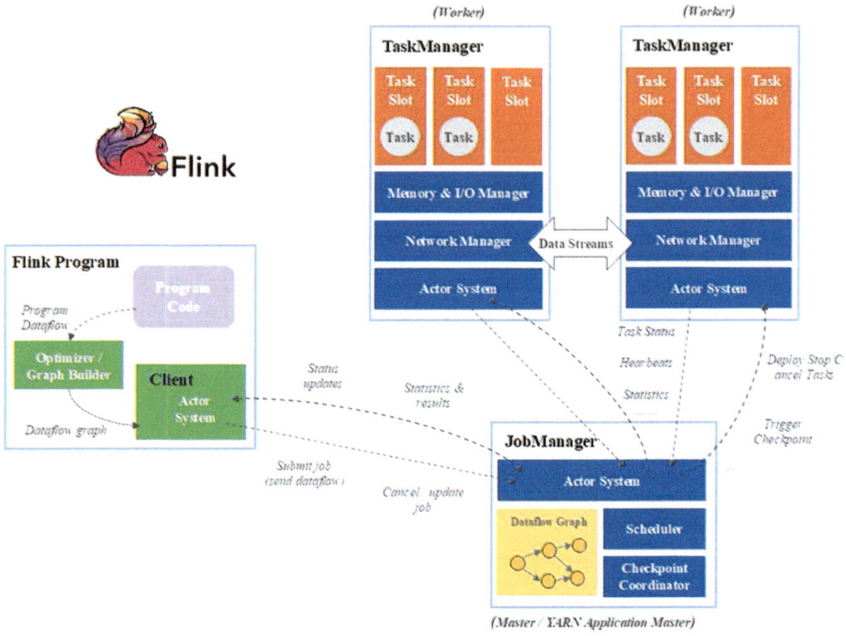

Fig. 9.5 Execution architecture diagram of Apache Flink

First, some regression models are deployed on the Apache Spark platform to quantify their prediction accuracy within a distributed computing environment.

Second, computations with varying batch sizes in the PHM system are simulated to assess the scalability of the real-time big data streaming architecture, particularly in terms of processing speed.

Applying the Distributed Computing Processing (DCP) strategy on the Spark platform enhances computational efficiency by dividing and parallelizing PHM big data. The DCP strategy process in this section follows these steps:

(1) A master–slave computing environment is configured on a Linux/Ubuntu system, with a Master name node and Worker data nodes designated. By distributing the two types of HDFS nodes across separate physical machines, parallel data computation is achieved.
(2) PHM big data input is converted into a parallelized Resilient Distributed Dataset (RDD) abstraction on the Spark platform, enabling data segmentation and manipulation.
(3) Computational resources are allocated to each data node, and distributed computing is performed on the parallelized RDD groups.
(4) Outputs of different parallelized RDD groups are combined in the original order to obtain the final computational results.

In this case study, the Extended Range Electric Vehicle Battery Life dataset from Chap. 8 (1,999,316 under CLTC standard/2,266,316 under NEDC standard) is selected for experimentation. The dataset includes 16 variables: time, voltage, current, charge, state of charge, maximum power output, maximum power input, electrical power, power loss, total output charge, total input charge, total output energy, total input energy, source voltage, internal resistance, and state of health. Additional relevant information is incorporated to predict the variables voltage and current.

(1) *Comparison of prediction accuracy*

Experiment 1 aims to evaluate the predictive performance of 8 different machine learning models on the CLTC and NEDC standard datasets within the Spark big data framework. Modeling for these 8 models is implemented using Apache Spark's PySpark-MLlib library alongside an external distributed computing framework. The models include RF, Linear Regression (LR), Decision Tree Regression (DT), Gradient-Boosted Trees (GBT), SVM, ELM, K-Nearest Neighbors (KNN), and MLP. These machine-learning models vary in topology, operational principles, and applicable scenarios. The model prediction performance across different datasets is effectively evaluated using four deterministic metrics: MAE, RMSE, P) and Kling-Gupta Efficiency (KGE). Table 9.2 presents the results of the prediction error metrics for the eight comparison models on the CLTC and NEDC datasets.

Figure 9.6 presents the error stacking chart for each model. Figure 9.7 shows the scatter plot of predicted versus actual values for each model.

Based on Table 9.2 and Figs. 9.6 and 9.7, the following conclusions can be drawn:

The topology and characteristics of each model significantly impact its performance under the Spark distributed computing framework. DCPRF and DCPKNN show the best performance on the CLTC dataset, particularly in terms of MAE and RMSE for voltage and current predictions, indicating high adaptability and stability on the Spark platform for this data. DCPELM exhibits the highest error, likely due to its sensitivity to data distribution, resulting in poor fitting for complex data. As dataset size and complexity increase, the nonlinear models DCPRF and DCPGBT demonstrate better adaptability and predictive performance, while the simpler DCPLR performs poorly, underscoring the importance of selecting suitable data-driven models for large datasets and complex patterns. Regarding correlation metrics P and KGE, DCPRF, DCPKNN, and DCPMLP perform well, with correlation coefficients close to 1, suggesting that these models not only predict accurately but also effectively capture the relational features within the data.

This experiment shows that the Spark-based distributed computing architecture effectively supports predictive tasks for various machine learning models in big data environments. DCPRF and DCPKNN exhibit optimal performance on this dataset, achieving low error rates and high predictive accuracy. In contrast, DCPELM performs poorly in this task, making it unsuitable for voltage and current predictions on this dataset. Considering error, correlation, and adaptability to distributed computing, DCPRF and DCPKNN are the recommended models under the conditions of this experiment.

Table 9.2 Predictive error and metrics for models on CLTC and NEDC datasets

Dataset	Model (on spark)	Variable	Accuracy performance			
			MAE	RMSE	P	KGE
CLTC standard	DCPRF	Voltage	0.0982	0.2863	0.9998	0.9982
		Current	0.0724	0.2930	1.0000	0.9991
	DCPLR	Voltage	0.6068	1.4651	0.9935	0.9930
		Current	0.4701	1.0313	0.9996	0.9942
	DCPDT	Voltage	0.1632	0.4108	0.9995	0.9981
		Current	0.1319	0.3854	0.9999	0.9984
	DCPGBT	Voltage	0.4396	0.6869	0.9986	0.9948
		Current	0.3451	0.5739	0.9999	0.9889
	DCPSVM	Voltage	0.6135	0.8777	0.9978	0.9927
		Current	1.9770	2.2820	0.9987	0.1023
	DCPELM	Voltage	1.4079	2.1312	0.9862	0.9830
		Current	4.2928	6.6746	0.9840	0.6398
	DCPKNN	Voltage	0.1046	0.2579	0.9998	0.9994
		Current	0.4137	0.9000	0.9997	0.9941
	DCPMLP	Voltage	0.1443	0.6348	0.9988	0.9959
		Current	0.2138	0.7405	0.9998	0.9798
NEDC standard	DCPRF	Voltage	0.1177	1.7221	0.9944	0.9789
		Current	0.1185	3.6073	0.9948	0.9834
	DCPLR	Voltage	0.9761	4.1500	0.9668	0.9319
		Current	0.5935	1.2064	0.9994	0.9943
	DCPDT	Voltage	0.1507	0.4658	0.9996	0.9988
		Current	0.1162	0.5021	0.9999	0.9984
	DCPGBT	Voltage	0.5220	1.7060	0.9945	0.9820
		Current	0.4365	6.4362	0.9837	0.9642
	DCPSVM	Voltage	0.7058	4.1659	0.9662	0.9416
		Current	1.8053	2.1629	0.9987	0.7053
	DCPELM	Voltage	1.8040	4.6099	0.9585	0.9425
		Current	4.3395	6.1755	0.9849	0.9755
	DCPKNN	Voltage	0.1407	3.2061	0.9805	0.9490
		Current	0.3656	1.3869	0.9992	0.9894
	DCPMLP	Voltage	0.2183	3.9627	0.9695	0.9442
		Current	0.1794	0.2655	1.0000	0.9943

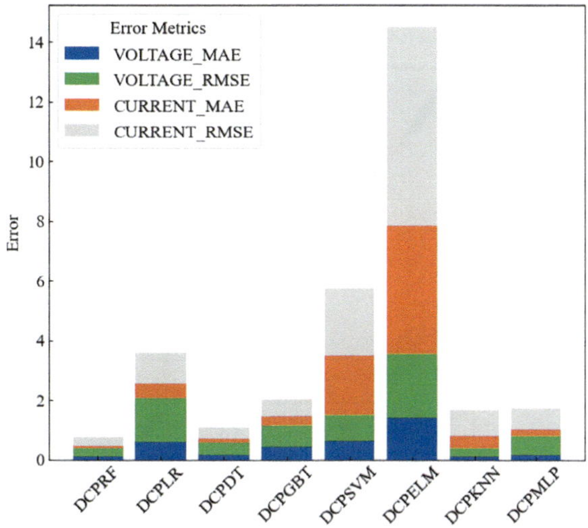

Fig. 9.6 Stacking graphs of prediction errors of each model on CLTC dataset

(2) *Comparison of forecast expansibility*

Experiment 2 primarily evaluates the computational time efficiency and scalability of the PHM big data streaming architecture under different models, batch sizes, and core configurations. The experiment is conducted on a Spark cluster consisting of four identical physical machines, each equipped with an Intel Core i7-9700 3.0GHz CPU with 8 cores and 12 execution threads, 16GB RAM, and an NVIDIA GeForce RTX 2060 GPU. The machines run Ubuntu 20.04.6 LTS (64-bit version). To simulate datasets of varying batch sizes, data replication and augmentation are used, with batch sizes ranging from 1 to 200 GB. Each model is tested under 200 experimental runs with combinations of batch sizes (1 GB, 5 GB, 10 GB, 50 GB, 100 GB, 150 GB, and 200 GB) and core numbers (2, 4, 8, 24, and 48 cores) to assess the performance of the distributed computing architecture.

Figure 9.8 presents the scalability performance results for each model in distributed computing. The horizontal axis represents the batch data size (in GB), and the vertical axis represents computation time (in seconds). A point (s, t) indicates that the respective big data streaming prediction model requires t seconds to process a batch of data of size s GB.

By analyzing the chart, it can be seen that different machine learning models significantly increase the computation time when the data size increases. However, the effect of increasing the number of cores on reducing the calculation time is different. The following is the specific performance of different models in the experiment:

(1) Fig. 9.8 shows that the calculation time of all models increases linearly with the increase of data batch size, indicating that the calculation cost also increases proportionally with the increase of data volume. This linear scaling feature validates the scalability of Spark clusters in processing large-scale data, especially

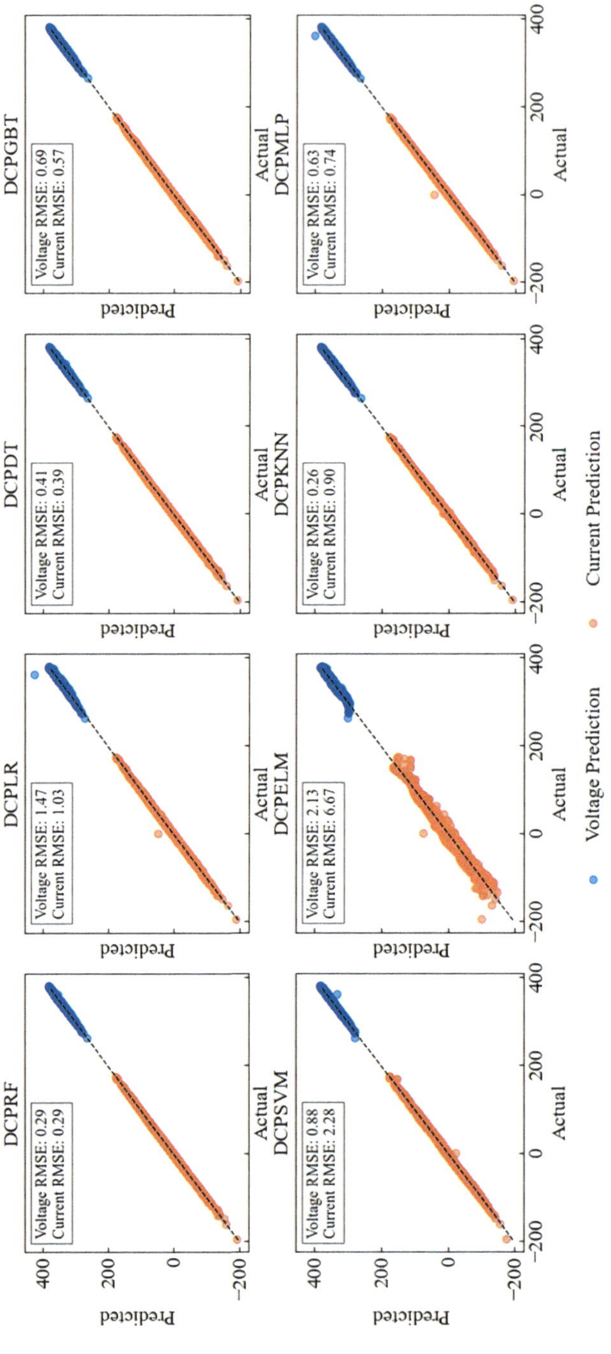

Fig. 9.7 Scatter comparison diagram of predicted values and actual values of each model in CLTC dataset

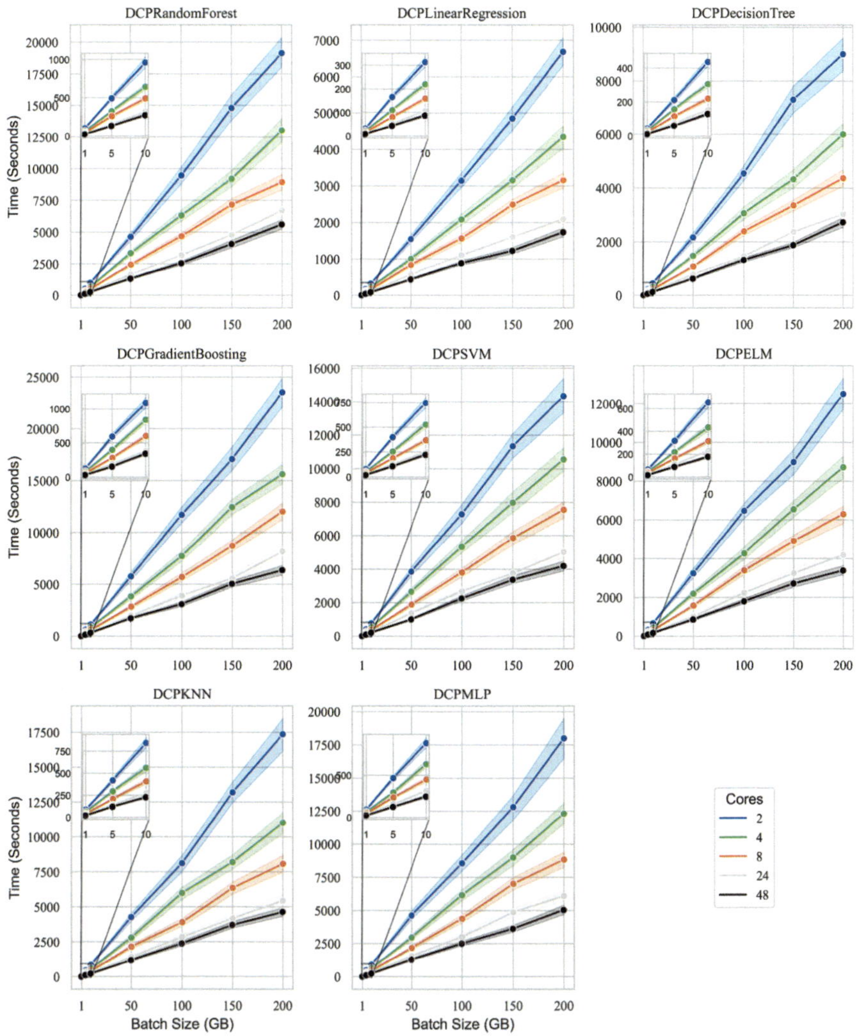

Fig. 9.8 Calculation time of each model in Spark cluster changes with batch size and number of cores

in large data stream batch processing can maintain reasonable computational efficiency.

(2) Under different core number configurations, the execution schedule of each model shows obvious differences. Overall, increasing the number of cores (from 2 to 48) significantly reduced the computational time for each model. This trend is particularly evident in the case of large data volumes, demonstrating the parallel computing capabilities of Spark's distributed architecture in a multi-core environment. When the number of cores reaches a certain level (such as 48

cores), the decline of calculation time gradually decreases, indicating that the marginal effect of increasing the number of cores decreases in some cases.

(3) From the specific model, the computing time of DCPRF and DCPMLP is relatively long in the big data batch processing and multi-core configuration, while the computing time of DCPLR and DCPELM is relatively short. This is closely related to the computational complexity of the model and how well Spark optimizes the algorithm. DCPRF and DCPMLP have high model complexity and computational requirements, so the time is more significant in big data and high-core configurations. The calculation time of DCPSVM and DCPGBT is also relatively long in large data batches, but the reduction is more obvious when the number of cores increases. This shows that such algorithms can benefit significantly from the allocation of multi-core resources in the process of parallelization.

(4) Overall, the experimental results show that the Spark cluster has good scalability under the architecture of processing PHM big data streams. Especially for computationally intensive models such as DCPRF, DCPSVM, and DCPGBT, multi-core configurations can significantly reduce computation time. This suggests that the PHM architecture can leverage Spark's distributed nature to effectively address the processing needs of large data streams by increasing the number of cores. However, some models (such as DCPLR and DCPKNN) show a more gradual decline in computation time at a given number of cores, indicating that such models may have a parallelization bottleneck when further increasing the number of cores. Therefore, in practical applications, choosing the appropriate core configuration according to the model can obtain higher computational efficiency.

Experiment 2 verifies the scalability of PHM big data flow architecture on the Spark platform, and the configuration of different number of cores and data batch sizes has a significant impact on computing efficiency. The increase of the number of cores can effectively reduce the calculation time of most models, but the response of different models to the increase of the number of cores is different due to the computational complexity. In practical application, reasonable allocation should be made according to the computing characteristics, data batch size, and hardware resources of the model to achieve the optimal balance of computing time and resource utilization.

9.3.3 Challenges and Analysis

Although big data technology has broad application prospects in PHM system, it still faces many challenges in the practical application process. For the problems of data heterogeneity, high-dimensional data, real-time, data quality and data storage, it is necessary to combine advanced data processing technology, computing architecture, and data management strategy to optimize. The solution of these challenges not only depends on the innovation of technical means, but also requires a deep understanding of the business requirements of PHM systems in order to develop appropriate solutions. Overcoming these problems will help promote the wider application of big data in PHM systems and significantly improve the system's forecasting and decision-making capabilities.

(1) Data heterogeneity

In PHM system, data usually comes from different sensors, devices, systems, and even different platforms, and these data may have differences in format, structure, acquisition frequency, accuracy, and semantics. This heterogeneity increases the difficulty of data fusion and integration. To solve this problem, data fusion technology needs to be applied to unify and standardize data structures to ensure that data from different sources can effectively collaborate in the same PHM system. Data fusion requires not only efficient technical means, but also a deep understanding of data semantics in order to eliminate differences between data sources. In addition, data cleaning and pre-processing are also essential in this process to ensure data integrity and consistency.

(2) High dimensional data processing

The sensor data commonly found in PHM systems have high dimensional characteristics, and these data may contain hundreds or even thousands of features. High dimensional data will bring "dimensional disaster" in modeling and calculation, which not only increases the complexity of calculation, but also may lead to overfitting of the model and reduce the accuracy. Dimensionality reduction and feature selection techniques are often used to reduce data complexity, enhancing model efficiency and generalization. However, a key challenge lies in preserving the most important information during this reduction process. Finding suitable feature selection algorithm and balancing calculation cost and prediction accuracy are the key problems that need to be solved in high dimensional data processing of the PHM system.

(3) Real-time requirements

Many PHM applications, particularly failure prediction and health management, require real-time monitoring and analysis of data to detect and respond to anomalies in a timely manner. The traditional cloud computing architecture has a certain delay in processing big data, which is not suitable for scenarios with strict real-time requirements. To improve real-time performance, PHM system can combine edge computing and cloud computing, and reasonably allocate computing tasks to edge nodes and cloud. Edge computing can quickly process data near the data source and reduce the delay of data transmission, which is suitable for real-time monitoring and preliminary analysis. The cloud allows for more complex processing and long-term storage to support in-depth trend analysis and historical data modeling. However, the combination of edge computing and cloud computing also faces the challenges of data synchronization, communication delay and resource allocation. Figuring out how to ensure effective collaboration between the two is a pressing issue that needs to be addressed.

9.4 Conclusions

This chapter focuses on the application of big data in PHM system and analyzes the key role of big data storage and distributed prediction models in improving the performance of PHM system combined with specific experiments. The experimental results show that by building a robust distributed computing platform (such as Spark cluster) and comparing different machine learning models, the computational efficiency of the PHM system can be significantly improved when processing large-scale data. Specifically, the performance of different models in distributed environments reveals their applicable scenarios and limitations in PHM systems, providing empirical evidence for the optimization of big data platforms and model selection.

It can be seen from the experiment that DCPRF and DCPKNN models have excellent performance on large data sets, which can ensure the prediction accuracy while maintaining high computational efficiency. However, the processing efficiency and accuracy of other models such as DCPSVM and DCPELM on high-dimensional data or large-scale data still need to be improved. Through the analysis of experimental data and error evaluation, we can see the prediction ability of different models in the big data environment, which provides strong support for the construction of a stable and efficient PHM system.

In the future research direction, further integration of AI-driven advanced analytics technologies (such as adaptive learning algorithms) into big data platforms will help enhance the intelligence and autonomy of PHM systems. Meanwhile, by combining edge computing with cloud computing, the data transmission delay in the PHM system can be effectively reduced, so as to better meet the needs of real-time monitoring and prediction. Future PHM systems can be explored based on edge-to-cloud collaborative architecture to achieve real-time condition monitoring and failure prediction of large-scale industrial equipment.

References

1. An XY, Huang ZY, Dou ZF, Lu FL, Wang Q (2025) Predicting aging of IGBT solder layer using saturation voltage approach with CPO-SVR data modeling. Microelectron Reliab 164:115553. https://doi.org/10.1016/j.microrel.2024.115553
2. Celaya JR, Kulkarni CS, Biswas G, Goebel K (2013) Towards a model-based prognostics methodology for electrolytic capacitors: A case study based on electrical overstress accelerated aging. International Journal of Prognostics Health Management 3:33
3. Celaya J. R., Saxena A., Wysocki P., Saha S., Goebel K. (2010) Towards prognostics of power MOSFETs: Accelerated aging and precursors of failure. Annual Conference of the PHM Society 2 (1)
4. Chen GQ, Hao R, Yang M (2025) Popularity-based multiple-replica cloud storage integrity auditing for big data. Futur Gener Comput Syst 163:107534. https://doi.org/10.1016/j.future.2024.107534
5. Fan HW, Ren ZF, Zhang XH, Cao XG, Ma HW, Huang JX (2024) A gray texture image data-driven intelligent fault diagnosis method of induction motor rotor-bearing system under

variable load conditions. Measurement 233:114742. https://doi.org/10.1016/j.measurement.2024.114742
6. He Q, Chen YT, Dong QH, Wang Y (2020) Mining moving object gathering pattern based on Resilient Distributed Datasets and R-tree index. Neurocomputing 393:194–202. https://doi.org/10.1016/j.neucom.2018.09.107
7. Huang C, Bu SQ, Lee HH, Chan CH, Kong SW, Yung WKC (2024) Prognostics and health management for predictive maintenance: A review. J Manuf Syst 75:78–101. https://doi.org/10.1016/j.jmsy.2024.05.021
8. Kumar P, Raouf I, Kim HS (2023) Review on prognostics and health management in smart factory: From conventional to deep learning perspectives. Eng Appl Artif Intell 126:107126. https://doi.org/10.1016/j.engappai.2023.107126
9. Li HJ, Li JL, Duan XL, Xia JL (2025) Energy-aware scheduling and two-tier coordinated load balancing for streaming applications in apache flink. Futur Gener Comput Syst 166:107681. https://doi.org/10.1016/j.future.2024.107681
10. Lim S, Park D (2024) Improving Hadoop MapReduce performance on heterogeneous single board computer clusters. Futur Gener Comput Syst 160:752–766. https://doi.org/10.1016/j.future.2024.06.025
11. Lin ZH (2024) Data-driven solutions: Uncovering the hidden potential of big data technologies in in building low-carbon cities. Comput Ind Eng 197:110543. https://doi.org/10.1016/j.cie.2024.110543
12. Liu CJ, Wang XF, Jiang XM, Sui YF, Cao SD, Qi C, Liu HT (2024) A parallel and multi-scale probabilistic temporal convolutional neural networks for forecasting the key monitoring parameters of gas turbine. Eng Appl Artif Intell 138:109400. https://doi.org/10.1016/j.engappai.2024.109400
13. Lu QM (2024) Application of Data Visualization Technology in Fault Diagnosis and Maintenance of Intelligent Electromechanical Systems. Procedia Computer Science 243:716–723. https://doi.org/10.1016/j.procs.2024.09.086
14. Mavridis I, Karatza H (2017) Performance evaluation of cloud-based log file analysis with Apache Hadoop and Apache Spark. J Syst Softw 125:133–151. https://doi.org/10.1016/j.jss.2016.11.037
15. Mekki K, Derigent W, Rondeau E, Thomas A (2017) Data Lifecycle Management in Smart Building using Wireless Sensors Networks. IFAC-PapersOnLine 50(1):12944–12949. https://doi.org/10.1016/j.ifacol.2017.08.1796
16. Nectoux P., Gouriveau R., Medjaher K., Ramasso E., Chebel-Morello B., Zerhouni N., Varnier C. PRONOSTIA: An experimental platform for bearings accelerated degradation tests. In: IEEE International Conference on Prognostics and Health Management, PHM'12., 2012. IEEE Catalog Number: CPF12PHM-CDR, pp 1–8
17. Polato I, Ré R, Goldman A, Kon F (2014) A comprehensive view of Hadoop research—A systematic literature review. J Netw Comput Appl 46:1–25. https://doi.org/10.1016/j.jnca.2014.07.022
18. Qin LY, Wang XL, Yin LZ, Jiang ZH (2024) A distributed evolutionary based instance selection algorithm for big data using Apache Spark. Appl Soft Comput 159:111638. https://doi.org/10.1016/j.asoc.2024.111638
19. Quick D., Martini B., Choo K. R. (2014) Chapter 5 - Google Drive: Forensic Analysis of Cloud Storage Data Remnants. In: Quick D, Martini B, Choo K-KR (eds) Cloud Storage Forensics. Syngress, Boston, pp 95–126. https://doi.org/10.1016/B978-0-12-419970-5.00005-3
20. Rodríguez M, Tobón DP, Múnera D (2025) A framework for anomaly classification in Industrial Internet of Things systems. Internet of Things 29:101446. https://doi.org/10.1016/j.iot.2024.101446
21. Sadeghi-Nasab A, Ghaffarian H, Rahmani M (2023) Apache Flink and clustering-based framework for fast anonymization of IoT stream data. Intelligent Systems with Applications 20:200267. https://doi.org/10.1016/j.iswa.2023.200267
22. Shi HX, Geng J, Demichela M (2024) How does Prognostics and Health Management (PHM) develop in the engineering domain: A bibliometric analysis with critical review. Heliyon 10(24):e40920. https://doi.org/10.1016/j.heliyon.2024.e40920

23. Singh T, Gupta S, Satakshi KM (2023) Performance Analysis and Deployment of Partitioning Strategies in Apache Spark. Procedia Computer Science 218:594–603. https://doi.org/10.1016/j.procs.2023.01.041
24. Smith WA, Randall RB (2015) Rolling element bearing diagnostics using the Case Western Reserve University data: A benchmark study. Mech Syst Sig Process 64:100–131
25. Sonnenfeld G., Goebel K., Celaya J. R. (2008) An agile accelerated aging, characterization and scenario simulation system for gate controlled power transistors. Paper presented at the 2008 IEEE AUTOTESTCON,
26. Soori M., Jough F. K. G., Dastres R., Arezoo B. (2024) Blockchains for industrial Internet of Things in sustainable supply chain management of industry 4.0, a review. Sustainable Manufacturing and Service Economics 3:100026. https://doi.org/10.1016/j.smse.2024.100026
27. Rehman MH, Yaqoob I, Salah K, Imran M, Jayaraman PP, Perera C (2019) The role of big data analytics in industrial Internet of Things. Futur Gener Comput Syst 99:247–259. https://doi.org/10.1016/j.future.2019.04.020
28. Vini MJ, Nallathamby R, Robin CRR (2015) A Novel Approach for Replica Synchronization in Hadoop Distributed File Systems. Procedia Computer Science 50:590–595. https://doi.org/10.1016/j.procs.2015.04.090
29. Wang JL, Xu CQ, Zhang J, Zhong R (2022) Big data analytics for intelligent manufacturing systems: A review. J Manuf Syst 62:738–752. https://doi.org/10.1016/j.jmsy.2021.03.005
30. Wang LJ, Li XY, Wang CG, Ma X, Miao B, Song XN, Cheng RX (2024) Enhancing collaborative PHM of complex equipment: A MBSE approach for PEMFC health management systems. Int J Hydrogen Energy 94:1361–1373. https://doi.org/10.1016/j.ijhydene.2024.11.208
31. Younan MN, Houssein EH, Elhoseny M, Ali AA (2020) Challenges and recommended technologies for the industrial internet of things: A comprehensive review. Measurement 151:107198. https://doi.org/10.1016/j.measurement.2019.107198
32. Zhao YF, Li ZH, Lv SY (2024) Enhancing AI System Privacy: An Automatic Tool for Achieving GDPR Compliance in NoSQL Databases. Computers, Materials and Continua 80(1):217–234. https://doi.org/10.32604/cmc.2024.052310